DIGITAL ELECTRONICS

JOHN C. MORRIS

BA, IEng, FIEIE, CertEd

Senior Lecturer
Department of Electronics and Engineering Technology
Havering College of Further and Higher Education

Edward Arnold
A division of Hodder & Stoughton
LONDON   MELBOURNE   AUCKLAND

For Ian and Adam

'Still round the corner there may wait,
a new road, or a secret gate.'

J. R. R. Tolkien

© 1992 John C. Morris

First published in Great Britain 1992

Distributed in the USA by Routledge, Chapman and Hall, Inc.
29 West 35th Street, New York, NY 10001

British Library Cataloguing in Publication Data

Morris, John C.
  Digital Electronics
  I. Title
  621.3815

  ISBN 0–340–55638–2

Typeset in 10/11pt Palatino by Wearset, Boldon, Tyne and Wear
Printed and bound in Great Britain for Edward Arnold, a division
of Hodder and Stoughton Limited, Mill Road, Dunton Green,
Sevenoaks, Kent TN13 2YA by Thomson Litho Ltd., East Kilbride,
Scotland

Contents

Preface

This book is intended for readers with an understanding of basic electronic techniques who wish to develop their knowledge of the various digital techniques and applications that today increasingly play a major part in the world of practical electronics.

A discovery-based approach is used throughout that introduces the reader to pulse waveforms and generators before dealing with the types of logic families and gates that are available. Combinational and sequential logic networks are covered in detail showing how individual CMOS and TTL gates may be used, and where purpose built integrated circuits can be used to simplify a design. Display devices together with analogue-to-digital and digital-to-analogue conversion is included. The final chapter shows how appropriate instruments can be used to diagnose faults on digital circuits.

Student-centred methods are used throughout that employ manufacturers' data sheets and self test questions to reinforce the theory. Digital electronics is essentially a 'hands on' subject so, to help forge the all important link between knowing and doing, 25 practical investigations are included that allow the reader to verify theoretical concepts using the minimum amount of equipment.

The material covers the BTEC Digital Electronics NIII syllabus while sharing common ground with City and Guilds 224, 271 and GCSE courses. Each chapter includes a review section to enable the reader to recap the important points without having to hunt through the text. Background information if required can be found in the companion volumes *Electronics — Practical Applications and Design* (Morris, 1989) and *Analogue Electronics* (Morris, 1991).

It is my hope that the step-by-step methods adopted will make this book a source of interest to enthusiasts, technicians and teachers as well as the student of electronics.

The combined efforts of a number of people are reflected within these pages. I would like therefore to acknowledge R.S. Components Ltd. for permission to reproduce extracts from their current catalogue and data sheets. My colleagues, as always, have been kind and supportive, particularly Neville James whose attention to detail and helpful suggestions have led to many real improvements. Thanks also to Kevin Hallam for proof-reading the final draft and to Mike Lenard for taking the photographs.

Finally I should like to offer a special thank you to my family; to Ian and Adam, for their patience and understanding and to my wife Lin for her constant encouragement, help and skilled typing of the manuscript.

John C. Morris
Billericay, Essex. October 1991

Introduction

The information in this book is presented in a sequence that makes it suitable for use as a programme of study. Each topic however is separate and self contained and so may be studied in isolation if required.

The practical investigations are designed to allow you to prove to yourself the operation of logic circuits and I recommend that you attempt these when suggested by the associated text. A minimum of equipment is required and with one or two exceptions the investigations can be carried out in the home.

As you will discover many integrated circuits (ICs) or 'chips' are available in TTL or CMOS form. To enable either type to be used the power supply for most of the investigations is +5 V. There are a number of commercial logic tutors on the market that will allow all the practical investigations in this book to be constructed quickly and easily. However as these are quite costly, an alternative is to use the readily available 'breadboards' sold for the electronics constructor and experimenter. Circuits can be built easily and modified using this medium, with the advantage that all the components can be reclaimed and re-used.

The majority of logic circuits require binary input signals having logic levels '0' (0 V) and '1' (+5V). To supply these all that is required are simple toggle switches wired as shown in Fig. (i). For convenience dual-in-line (dil) switches can be used that have the same packaging as integrated circuits and will wire directly on to the board.

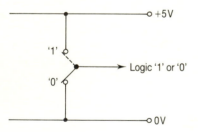

Fig. (i) Logic switch

To detect the logic state of a circuit light emitting diodes (LEDs) can be used as indicators; remember however that the current through a LED must be limited in order to prevent it burning out. A 470 ohm resistor placed in series with the diode will ensure safe operation. LEDs like switches are available singly or as dual-in-line packages containing a number of devices. The switches and indicators should be built as permanent features on the breadboard but, to give sufficient working space, it is a good idea to use a couple of breadboards here. The switches and indicators can be kept on one board leaving the other completely empty for use as a logic circuit construction site. One typical breadboard set up is shown in Fig. (ii).

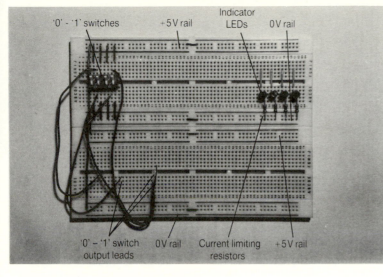

Fig. (ii) Logic breadboard arrangement

The 555 Timer

The 555 Universal Timer integrated circuit shown in Fig. (iii) is an extremely useful device because, by employing only a few external components, a number of quite complex circuits can be quickly constructed.

Internally the chip itself is a combination or mixture of digital and analogue circuitry consisting of two operational amplifiers, an RS flip-flop and a single transitor. For our purposes however its make-up is unimportant, all we shall be doing is using the 555 as a convenient tool to provide the function we require, e.g. as a clock pulse generator.

Where it is appropriate the practical investigations employ the 555 Timer. Do not be put off by this! The timer is a very cheap device that is easy to use and reliable in operation. When performing these investigations remember it is very easy to wire a circuit incorrectly. Work slowly and methodically; check the pin connections of the IC using the data sheets at the back of the book; and *always*, double check your circuit *before* switching on.

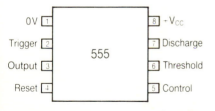

Fig. (iii) 555 Timer pin connections

1

Pulse Waveforms

Digital and analogue signals

Today the world of electronics is a mixture of analogue and digital circuitry. Many systems such as amplifiers, radios and televisions operate by using analogue signals. However, other systems such as watches, calculators and computers operate using digital signals. As time passes more and more traditional analogue systems are superseded by digital forms, notably, ni-cam digital stereo, the compact audio disc and audio cassette machines which employ digitally recorded magnetic tape. There is a significant difference between analogue waveforms or signals and those which are considered to be digital. It is a good starting point to establish clearly the difference between the two types of signal.

Analogue signals

The definition of such a waveform is one that has a continuously varying quantity (such as amplitude or frequency), with time. Two such signals are shown in Fig. 1.1(a) and 1.1(b). By studying these waveforms you may see that there are an infinite number of possible values between the minimum and maximum amplitudes.

Digital signals

A digital waveform can be divided into a finite number of levels as shown in Fig. 1.2(a), (b), (c). The waveform of Fig. 1.2(a) has seven possible levels (including 0 V) between the minimum and maximum amplitude, while those of Fig. 1.2(b) and 1.2(c) have only two levels. A morse

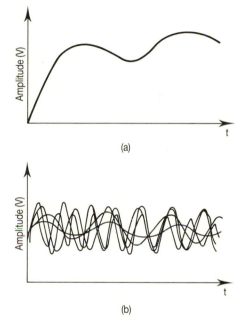

Fig. 1.1 Analogue waveforms

code signal has two levels, the information is conveyed by the 'on' time or duration of the pulses: a short duration pulse being a 'dot' and a longer one a 'dash'. The type of signal we are most concerned with however is the binary signal shown in Fig. 1.2(c) where there are two levels, the high or 'on' and the low or 'off', or as they are more commonly referred to, 'Logic 1' and 'Logic 0' respectively.

THOUGHT

So digital signals have specified levels and most digital systems respond to binary signals having logic levels '0' and '1'? This is true but we must now consider the important aspects of a 'pulse' or 'pulsating' waveform.

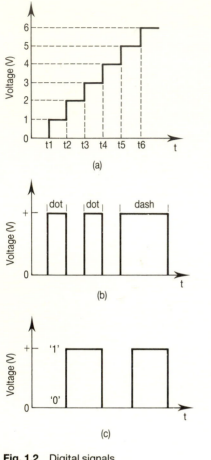

(a)

(b)

(c)

Fig. 1.2 Digital signals
(a)   Seven levels
(b)   Morse code
(c)   Binary code

Logic gates or digital circuits of any kind respond to or are triggered by pulses, either presented singly as shown in Fig. 1.3(a) or as a 'pulse train' in Fig. 1.3(b). You may see from these waveforms that there are two voltage levels involved; a low voltage that represents

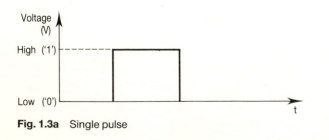

Fig. 1.3a   Single pulse

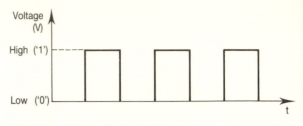

Fig. 1.3b   Pulse train

binary '0' and a higher voltage level representing binary '1'.

Pulse characteristics

Having established the levels for Logic '1' and '0' we must now look at the pulse itself. Fig. 1.4 shows an ideal pulse as it might be displayed on the screen of an oscilloscope. You can see that the pulse appears to change from Low to High instantaneously (in zero time) and that the corners are clean cut and square. The reality is somewhat different, Fig. 1.5 shows the pulse in greater detail. Here it may be seen that there is a finite time taken for the pulse to rise from Low to High, just as it takes time for it to fall from High back to Low, and note that these times may be different. Also the corners are likely to

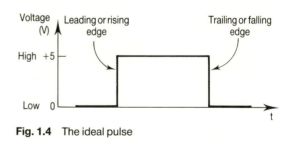

Fig. 1.4   The ideal pulse

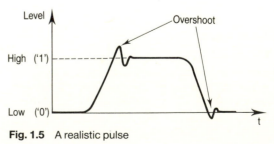

Fig. 1.5   A realistic pulse

be anything but square and will probably be curved with some 'ringing' or 'overshoot' present.

Logic devices respond fairly quickly so that the time a pulse takes to change level is important, as is the actual duration of the pulse itself. These characteristics are indicated in Fig. 1.6.

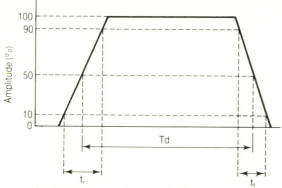

tr = 'rise time' of the pulse. This is the time that the pulse takes to rise from 10% to 90% of its final or maximum steady value.

tf = fall time of the pulse. The time taken for the pulse to fall from 90% to 10% of its maximum steady amplitude.

Td = pulse duration. The time interval measured at the 50% amplitude points on the waveform.

Fig. 1.6 Pulse characteristics

Pulse repetition frequency (prf)

This refers to the number of pulses that occur in one second, e.g. if the prf is 100 Hz then 100 pulses occur in 1 second. This is related to the Periodic Time (T) of the pulse.

$$\text{Periodic Time (T)} = \frac{1}{\text{prf}}$$

Note These definitions are similar to those you have come to know in relation to sine wave signals:

$$f = \frac{1}{T} \text{ Hz} \qquad T = \frac{1}{f} \text{ seconds}$$

There is a great difference where pulse waveforms are concerned, however, and this can be seen by studying the waveforms of Fig. 1.7. You may notice that in both cases the periodic time (T) and hence the frequency (f) is the same. But

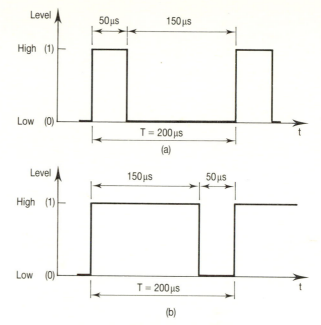

Fig. 1.7 Pulse repetition frequency waveforms

in Fig. 1.7(a) the pulse duration is 50 μs while the gap or 'space' between the pulses is three times as great at 150 μs. In Fig. 1.7(b) however the pulse is three times greater than the space between pulses. In both cases the frequency (f) is the same:

$$\therefore f = \frac{1}{T} = \frac{1}{200 \ \mu s} = \frac{1}{200 \times 10^{-6}} = 5 \text{ kHz}.$$

When considering pulse waveforms it is important that the duration of the pulse itself is quoted together with the time between pulses. This may be achieved in two ways.

The Mark to Space Ratio
With reference to Fig. 1.8, the 'on' time or pulse

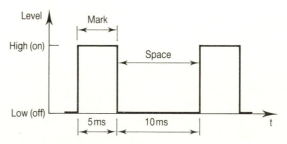

Fig. 1.8 Mark to space ratio

duration is termed the 'mark' time; and the 'off' time or space between pulses is termed the 'space' time. The ratio of pulse 'on' to pulse 'off' time is the mark to space ratio:

Mark to space ratio =

$$\frac{\text{time 'on'}}{\text{time 'off'}} = \frac{5 \text{ ms}}{10 \text{ ms}} = \frac{1}{2} \text{ or } 1:2$$

It follows that a waveform with a mark to space ratio of 1:1 has a pulse duration which is the same as the space between the pulses.

The Pulse duty cycle
This expresses the 'on' time of a pulse as a percentage of the complete periodic time, i.e. the amount of the combined mark and space time that is occupied by the pulse itself.

$$\text{Pulse duty cycle} = \frac{\text{time 'on'}}{\text{time 'on'} + \text{time 'off'}} \times 100\%$$

for the waveform of Fig. 1.8:

$$\text{Pulse duty cycle} = \frac{5 \text{ ms}}{5 \text{ ms} + 10 \text{ ms}}$$

$$= \frac{5}{15} \times 100\% = 33.3\%$$

From this you may see that a waveform with a pulse duty cycle of 50% must have a mark to space ratio of 1:1 since half the periodic time is occupied by the pulse.

Self Assessment 1

1   A pulse waveform has a duty cycle of 20% and a frequency of 2 kHz.
    Calculate (a) the pulse duration; (b) the space time; and (c) the mark to space ratio.
2   A 20 kHz pulse waveform has a mark to space ratio of 3:2. Calculate (a) the mark time; (b) the space time; and (c) the pulse duty cycle.

To help familiarize yourself with the practical pulse characteristics perform Practical Investigation 1.

A pulse waveform may contain noise and distortion that conspire to make it less than ideal. A variety of these problem pulse types are

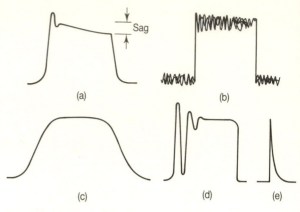

Fig. 1.9   Pulse types
(a)   With overshoot and sag
(b)   Noisy
(c)   Slow rise and fall times
(d)   'Ringing'
(e)   Very narrow

shown in Fig. 1.9. It is in our interest to provide a pulse which is as near to the ideal shape as possible, because only in this way can digital circuits operate reliably.

Producing a single 'clean' pulse

When building and testing digital circuits there is often a need to supply single pulses. The simplest way to do this is by using the switch arrangement of Fig. 1.10. By moving the switch from position 0 to position 1 and back to position 0 a pulse is created. Combinational logic circuits respond to the level or amplitude of the input signal, so this arrangement is perfectly adequate. If there are four inputs to a circuit, then four input switches are required to establish the input conditions. However, where sequential circuits such as counters and registers

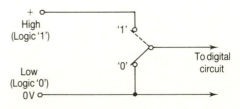

Fig. 1.10   Simple pulse switch

PRACTICAL INVESTIGATION *1*

Pulse Characteristics

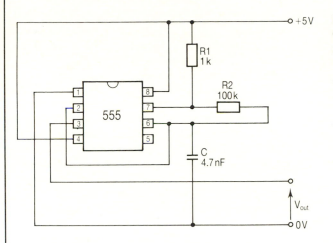

Equipment
Oscilloscope (5 MHz bandwidth minimum)
Power supply
555 Timer IC
Breadboard
1 k, 100 k resistors
4.7 nF 47 pF capacitors

Method
1. Build the circuit shown above initially using C = 4.7 nF.
2. Monitor the output from pin 3 with the CRO, and from the display determine:
   - (a) The periodic time (T).
   - (b) The mark time (T_{on}).
   - (c) The space time (T_{off}).
3. Examine the leading and trailing edges and try to estimate the rise and fall times (tr and tf).
4. Replace C and 4.7 nF with a 47 pF capacitor and repeat step 2.
5. Using the CRO controls display the leading edge and measure the rise time (tr).
6. From the falling edge establish the fall time (tf).
   Note For these measurements you may need to expand the CRO timebase using the × magnification facility.

Results
1. What is the mark to space ratio in each case?
2. Which pulse would be the best for triggering digital circuits and why?

are concerned there is a problem. Circuits of this type are pulse operated, if one pulse is required the input switch must produce just one single pulse. Within mechanical switches a phenomenon called 'contact bounce' occurs, when the actual switch contacts are closed they may open and shut a few times due to the spring action of the switch mechanics. This happens extremely quickly and is not evident to the switch operator. The result is a pulse that may look like the one shown in Fig. 1.11. The circuit

may interpret this as a number of pulses and will trigger accordingly. In the diagram shown

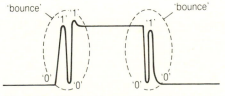

Fig. 1.11 'Contact bounce' waveform

the switch is moving from '0' to '1' and back to '0' giving one pulse. The bouncing contacts make it appear:– as 0101010 which is three such pulses. The solution is to use a bounce suppressor, or as it is more commonly known a 'debounced switch'. There are a number of circuits that will effectively debounce the output of a switch and two of these are shown in Figs 1.12 and 1.13.

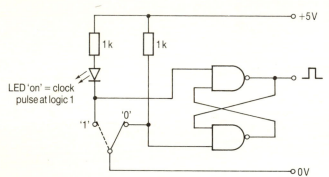

Fig. 1.12   Debounced switch using NAND gates

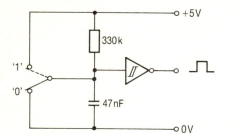

Fig. 1.13   Debounced switch using a Schmitt trigger

Note Some mechanical switches are debounced by coating the contacts with mercury, such types are called 'mercury wetted' and are available as reed relay inserts.

Before you start to investigate clocked logic circuits practically, you will have to ensure that the clock pulses are supplied from one of these circuits. More about this in Chapter 4.

Producing pulses with fast rise and fall times

There are many occasions when it is required to trigger a circuit with a waveform that has very

poor rise and fall times (such as the one in Fig. 1.9(c)). If this waveform was used there would be a fair chance that unreliable operation could result. What it needs is 'squaring up', and the circuit which does this is called the Schmitt trigger. Consider the waveform shown in Fig. 1.14; this circuit is a type of level sensing switch, where the output from the Schmitt trigger remains low until the input signal rises above a threshold voltage called the upper trigger voltage (UTV), when the output then goes high. It remains at this level until the input signal falls below the lower trigger voltage (LTV), when the output returns to the low level. Notice that the two threshold voltages are at different levels so that the circuit does possess some 'hysteresis', i.e. the later value will be dependent on the previous event, but in practice however this is unimportant. Schmitt triggers are today available as integrated circuits (Fig. 1.15(a), (b)).

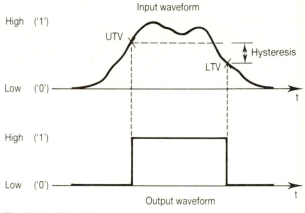

Fig. 1.14   Schmitt trigger waveform

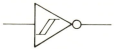

Fig. 1.15a   Schmitt trigger logic symbol

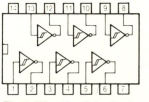
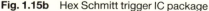

Fig. 1.15b   Hex Schmitt trigger IC package

PRACTICAL INVESTIGATION 2

The Schmitt Trigger

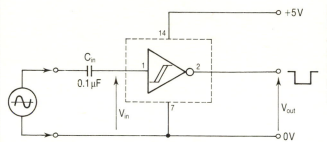

Equipment
Power supply (5 V)
40106, 7414 (Hex Schmitt triggers)
Signal generator
Breadboard
0.1 µF capacitor
Dual beam CRO

Method
1 Build the circuit shown above (there are six Schmitt triggers in this IC) making sure that pins 14 and 7 are connected to +5 V and 0 V respectively.
2 Set the signal generator to give a 1 kHz sine wave.
3 Monitor V_{in} and V_{out} using the CRO.
4 With $V_{in} = 0$ V measure the d.c. level of V_{out} using the CRO.
5 Increase the amplitude of V_{in} until V_{out} appears.
6 Measure the amplitude of V_{out} and note what happens as the amplitude of V_{in} is varied.
7 By aligning V_{in} and V_{out} estimate the upper and lower trigger voltages of the Schmitt trigger — to do this make sure you establish the 0 V level of V_{in}.

Results
1 When V_{in} is 0 V what is V_{out}?
2 When V_{in} is +5 V what is V_{out}?
3 If a positive output pulse is needed what additional circuitry would be required?

Perform the practical investigation of the Schmitt trigger and determine its characteristics.

So we are now in a position to generate a single pulse that is reasonably clean and noise free (it has flat peaks and troughs), with acceptably fast rise and fall times. Furthermore if we are presented with a pulse that has slow rise and fall times and unsuitable levels or polarity we can use a Schmitt trigger to create the desired pulse.

THOUGHT

What about a pulse that is too narrow like that shown in Fig. 1.9e? Occasionally there is a need for increasing the duration or stretching such a pulse so that it becomes wider but still rectangular. This we shall now consider.

Increasing the duration of a 'spiky' pulse

First let us see why some pulses are very sharp. There are situations and applications that require sharp trigger pulses, e.g. thyristor and triac circuits. However many perfectly respectable rectangular or square pulses can become spiky when they encounter a differentiating circuit, i.e. they have to pass through a capacitor. Fig. 1.16(a) shows differentiating circuits which consist of a capacitor (C) in series with a load (R). Notice in Fig. 16(a) how the single 1 ms pulse with an amplitude of +5 V becomes a positive and a negative spike. This subject is treated very thoroughly in electrical principles text books but briefly what happens is this. The fast rising edge A passes through the capacitor C to develop a positive voltage across R. During the 1 ms that the input is steady at +5 V the voltage across the capacitor increases towards +5 V and the output falls to 0 V. The fast trailing or falling edge B is a transition from +5 V to 0 V and this current passes through C to develop a negative voltage across R. Once the fall is complete the charge leaks away and the voltage returns to 0 V. In the circuit of Fig. 1.16(b) a similar thing happens, but the d.c. level starts at +5 V rather than 0 V with the result that the output will be positive, i.e. +5 V to +10 V and then to 0 V. One circuit that may be used to create a pulse with a specific duration is the Monostable.

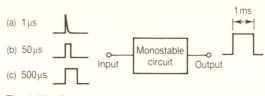

Fig. 1.17   The monostable

circuit simpler. The complete circuit is available as an IC (TTL 74122 or CMOS 4528) but it may be readily built using 2 NAND gates shown in Fig. 18(a) or a 555 Timer IC as shown in Fig. 1.19.

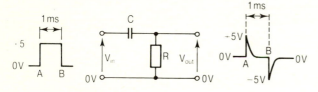

Fig. 1.16a   Differentiating circuit

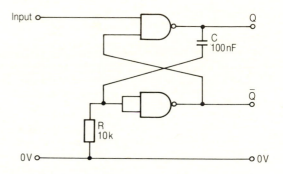

Fig. 1.18a   NAND gate monostable

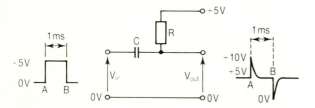

Fig. 1.16b   Differentiating circuit with a positive bias

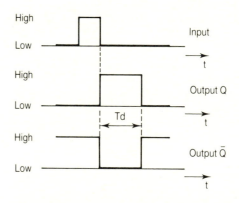

Fig. 1.18b   Monostable waveforms

The Monostable Circuit

This is a circuit that, when triggered by an input pulse, will produce an output pulse of a specified duration. You may see from Fig. 1.17 that the circuit provides an output pulse of 1 ms duration regardless of the duration of the input pulse.

A monostable can be designed to provide a pulse of any duration and can be constructed using discrete components or integrated circuits. It makes good sense to use integrated circuits (ICs) since this will make building the

NAND Gate Monostable Circuit

By studying the waveform diagrams Fig. 1.18(b) you may see that the monostable is triggered by the *falling* (or 'negative going') edge of the pulse and that this circuit gives a choice of output pulse, i.e. a positive pulse at Q that remains high for a specified period or a negative pulse at Q̄ that is the inverse of Q and remains low for a specified period.

The time constant CR determines the duration (Td) of output pulse and for the circuit shown

$Td = C \times R$; if $C = 100$ nF $R = 10$ k
    then:
$Td = 100 \times 10^{-9} \times 10 \times 10^3 = 1$ ms

The circuit shown in Fig. 1.19 uses the 555 Timer IC (see introduction) as a monostable. In this circuit the timing components are C and R with the duration of the output pulse Td given by 1.1 CR.

THOUGHT

Why is there a capacitor (Cin) and resistor (R1) connected at the input? Interesting one this: the 555 is triggered by a negative going pulse! The input pulse is positive so to ensure that the 555 receives a fast negative going signal a differentiator is connected between the input and the +5 V rail, and this will provide a negative pulse of 5 V from the positive input pulses.

Practical Investigation 3 allows you to study this monostable fully.

Producing a 'train' or row of Clock Pulses

We have defined the need for clean bounce free single pulses of a specified level and duration, but there is often a demand for a train of pulses of a specified frequency. Such signals provide clocking pulses for counting or 'gating' digital circuits. Purpose built pulse generators are available but it is quite convenient to build an astable multivibrator to suit particular circuit requirements. Practical Investigation 1 used such a circuit involving a 555 Timer IC. Let us explore this circuit further.

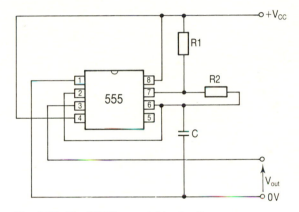

Fig. 1.20 The 555 Timer astable circuit

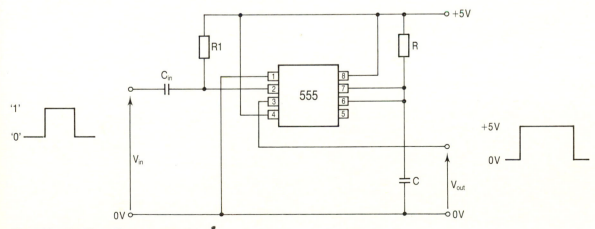

Fig. 1.19 555 Timer IC as a monostable

PRACTICAL INVESTIGATION 3

The Monostable

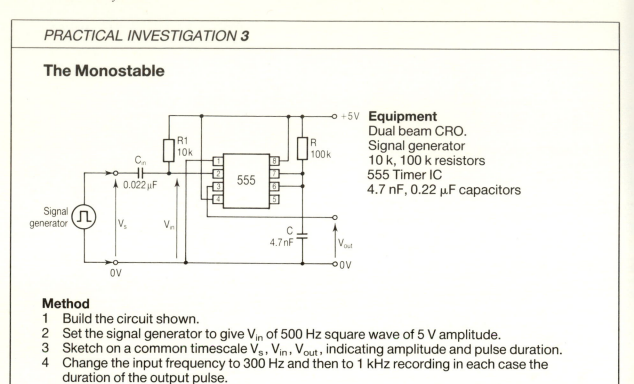

Equipment
Dual beam CRO.
Signal generator
10 k, 100 k resistors
555 Timer IC
4.7 nF, 0.22 µF capacitors

Method
1 Build the circuit shown.
2 Set the signal generator to give V_{in} of 500 Hz square wave of 5 V amplitude.
3 Sketch on a common timescale V_s, V_{in}, V_{out}, indicating amplitude and pulse duration.
4 Change the input frequency to 300 Hz and then to 1 kHz recording in each case the duration of the output pulse.

Results
1 Which type of pulse triggers the monostable?
2 Calculate the theoretical pulse duration provided by the monostable using $T_{on} = 1.1$ CR.
3 What effect does the input frequency have on the output pulse duration?

With this type of circuit not only must the pulse duration (T_{on}) be considered but also the gap between the pulses (T_{off}). This you may recall on page 3 when the mark to space ratio was discussed. In this case with the monostable the timing is achieved using a CR network. Here the 'on time' is determined by C and (R1 + R2) and the 'off time' by C and R2 and are defined by the following equations.

$$T_{on} \text{ (high)} = 0.69 \, (R1 + R2)C \text{ second}$$
$$T_{off} \text{ (low)} = 0.69 \, (R2C) \text{ second}$$

The periodic time (T) of the waveform is defined by

$$T = 0.69(R1 + 2 \times R2)C \text{ second}$$

Since the value of the capacitor is common to both T_{on} and T_{off}, the mark to space ratio can be determined using

$$M{:}S \text{ ratio} = \frac{T_{on}}{T_{off}} = \frac{R1 + R2}{R2}$$

Thus it is possible to design and build a pulse generator to provide a train of pulses with a

THOUGHT _____

How can a M:S of 1:1 be achieved using the 555? Well since M:S ratio = (R1 + R2)/R2 this appears impossible, but the secret is to make R2 much greater in value than R1, e.g. for R2 = 100 k and R1 = 1 k then the output will be virtually 1:1.

$$\frac{1\,k + 100\,k}{100\,k} = \frac{101}{100} = 1.01{:}1$$

i.e. A pulse duty cycle of approximately 50%.

PRACTICAL INVESTIGATION *4*

Pulse Generator

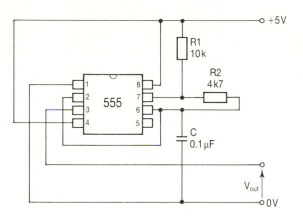

Equipment
Oscilloscope
555 timer
Power supply
Breadboard
4k7, 10 k resistors
0.1 µF capacitor

Method
1 Build the circuit shown above.
2 Using the CRO, monitor the output waveform, measure and record the pulse amplitude, periodic time and the mark and space times.
3 Calculate the periodic time (T), mark time (T_{on}) and space time (T_{off}) using the formulae:

$T_{on} = 0.69(R1 + R2)C$ second
$T_{off} = 0.69(R2)C$ second
$T = 0.69(R1 + 2 \times R2)C$ second
M:S ratio = (R1 + R2)/R2.

4 Repeat the above using R1 = 4k7, R2 = 10 k.

Results
1 Is there an agreement between each of the measured and the calculated values?
2 Which type of waveform would be the best for triggering or clocking digital circuits?

frequency, and mark to space ratio (pulse duty cycle) to a desired specification. Remember that when contemplating high frequency circuit design there will be inter-electrode, stray and other capacitances that will conspire to ruin the waveshape.

It is prudent to supply digital circuits with a clock pulse which is rectangular with a pulse duty cycle of 50% (M:S ratio 1:1). The 555 Timer IC is suitable for this, and for many of the investigations that follow it will be possible to use a 555 Timer that supplies low frequency pulses, i.e. has a pulse repetition frequency (prf)

of between 1 Hz and 2 Hz. In this way the circuit operation can be closely observed by slow clocking, in normal circuits however very fast clock speeds are often used.

Pulse Waveform Review

1 An analogue waveform is one that has an infinite number of possible levels.
2 A digital waveform has a finite or specific number of levels.

3   The most common form of digital signal is a binary signal that has two levels — '0' and '1' ('low' and 'high', or 'off' and 'on').

4   Digital circuits respond to logic levels: (Logic '0' and Logic '1').

5   An ideal pulse is clean and rectangular with defined levels.

6   A practical pulse may be noisy and distorted with overshoot and sag present, or it may be of a very short duration (spiky).

7   A pulse takes a finite time to change to another level. The rise time (tr) is the time taken for the voltage to rise from between 10% and 90% of its maximum amplitude. the fall time (tf) is the time taken for the voltage to fall from 90% to 10% of its total amplitude.

8   The duration of a pulse (Td) is the time interval measured between the 50% amplitude points on the waveform.

9   The pulse repetition frequency (prf) is the number of pulses occurring in one second.

10 The 'on' or 'high' time of a pulse is called its 'mark' time, while the 'off' or 'low' time is called its 'space' time.

11 The periodic time of a pulse is the time taken for one complete pulse cycle to occur, i.e. T = mark + space time. This is related to the prf. T = 1/prf second.

12 The ratio of pulse 'on' time to pulse 'off' time is called the 'mark to space ratio'

$$\text{M:S ratio} = \frac{\text{time 'on'}}{\text{time 'off'}}.$$

13 The pulse duty cycle expresses the 'on' time of a pulse as a percentage of the periodic time (T)

pulse duty cycle

$$= \frac{\text{time 'on'}}{\text{time 'on'} + \text{time 'off'}} \times 100\%.$$

14 Pulses used for sequential digital circuits need to be clean and noise free. When mechanical switches are used to supply single pulses they must be free of any contact bounce. This is achieved using a debouncing circuit that suppresses any spurious pulses.

15 A Schmitt trigger is a circuit that effectively 'squares' a pulse. This can be used to create a rectangular pulse from a distorted waveform.

16 A 555 Timer is an integrated circuit that contains digital and analogue circuitry. It is suitable for many pulse generating applications.

17 A monostable can be used to provide a pulse of specified duration whenever it is triggered. This can be used to lengthen or stretch a pulse that has a very short duration.

18 An astable is a pulse generator that can supply a train of clean pulses that are suitable for clocking or triggering digital circuits.

19 A suitable clock pulse generator for triggering digital circuits can be built using a 555 Timer IC as an astable.

Self Assessment Answers

Self Assessment 1

1   (a)   f = 2 kHz $T = 1/(2 \times 10^3) = 500\ \mu s$
            pulse duty cycle = 20%
            pulse duration = $0.2 \times 500\ \mu s = 100\ \mu s$
     (b)   space time = $500\ \mu s - 100\ \mu s = 400\ \mu s$
     (c)   mark to space ratio = 1:4.

2   F = 20 kHz $T = 1/(20 \times 10^3) = 50\ \mu s$
     mark to space ratio = 3:2
     the total periodic time occupies 3 + 2 = 5 units

$$1\ \text{unit} = \frac{50\ \mu s}{5} = 10\ \mu s$$

     (a)   The mark time = $3 \times 10\ \mu s = 30\ \mu s$
     (b)   The space time = $2 \times 10\ \mu s = 20\ \mu s$
     (c)   The pulse duty cycle

$$= \frac{30\ \mu s}{30\ \mu s + 20\ \mu s} \times 100\% = 60\%.$$

2

Logic Gates

'A logic gate is a device which can have more than one binary input but a single binary output. The state of the output is determined by the input conditions.'

Every logic gate can be depicted as a symbol, Boolean expression and truth table. While the truth tables and Boolean expressions are universally accepted there are considerable differences in the gate symbols. In the United Kingdom logic symbols are devised by the British Standards Institute (BSI). Unfortunately however, the American standard symbols (MIL/ANSI) for logic gates have a greater following. This has now reached such a level that the BSI symbols for logic gates are being almost universally

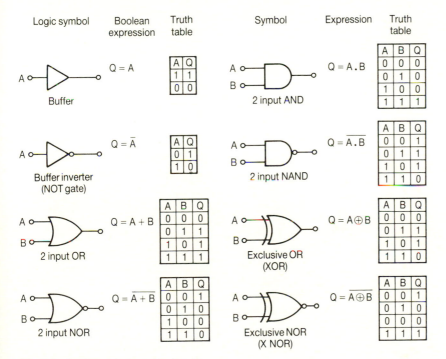

Fig. 2.1 Logic symbols and truth tables

superseded by the MIL/ANSI symbols. The reason for this is probably that the American versions are clearer and more easily understood. It is quite possible that in the not to distant future BSI symbols will be changed so that they conform to those of the MIL/ANSI. For reasons of clarity MIL symbols for logic circuits will be used throughout this book.

Fig. 2.1 shows the common logic gate symbols together with their Boolean expressions and truth tables.

Gate Operations

The gate operation can be represented by the Boolean expression and the truth table, this is summarized as follows.

The inverter or NOT gate

This produces an output which is the inverse or opposite of the input signal. Therefore if the input to an inverter is Logic 1 the output will be Logic 0.

The OR gate

A Logic 1 output is produced if any input is at Logic 1. For a gate with four inputs A, B, C, D the Boolean expression would be:

$Q = A + B + C + D$

with the plus sign (+) denoting the OR function, showing that the output $Q = '1'$ if any of the inputs A, B, C, D are at '1'.

The NOR gate

This is the inverse of the OR gate, i.e. it will give an output that is the opposite of that gate. The Boolean expression for a four input gate will be:

$Q = \overline{A + B + C + D}$.

The bar over the top indicates that the OR function is inverted, i.e. a NOT OR function, showing that if any is at '1' then $Q = '0'$.
Note Negating is a term sometimes used instead of inverting.

The AND gate

This gives a Logic 1 output when *all* the inputs are at Logic 1. The Boolean expression for a four input gate would be:

$Q = A.B.C.D$.

The dot (.) denotes the AND function.

The NAND gate

This is the inverse of the AND gate giving an output that is the opposite of the AND function. A four input gate may be expressed as:

$Q = \overline{A.B.C.D}$

showing that if any input is at '0', $Q = '1'$.

The two 'Exclusive' gates

The Exclusive OR (XOR) gate is like the OR gate except that it produces an output only if *one* of the two inputs is at Logic 1. If *both* are at Logic 1 the output will be Logic 0. The Boolean express is:

$Q = A \oplus B$.

This is a special gate for the following reasons:

1   It can only have two inputs.
2   The Boolean expression for exclusive OR $\oplus$ is not a real one! i.e. it is difficult to perform Boolean algebra using this expression.

As you might expect the Exclusive NOR (XNOR) is the inverse of the exclusive OR function and has the Boolean expression:

$Q = \overline{A \oplus B}$.

By studying the symbols you can see that inversion is represented by a circle shown at the output of the gate. As with the exclusive OR it is difficult to perform Boolean algebra using this expression.

THOUGHT _____

Is the output always represented by Q? No, virtually any term can be used. Q is common and so is F. So, Q = A.B or F = A.B or Z = A.B are all equally acceptable.

_____

So far we have dealt with binary pulses and logic gates, it is now time to look inside the integrated circuits themselves and identify the different logic families.

Logic Families: a first look

When logic gates started to be produced in integrated circuit form 'Resistor Transistor Logic' (RTL) circuitry was used, this was followed by 'Diode Transistor Logic' (DTL) and then 'Transistor Transistor Logic' (TTL). Other configurations (families) have since been developed, notably 'Complementary Metal Oxide Semiconductor' (CMOS) logic and 'Emitter Coupled Logic' (ECL). ECL is extremely fast in operation but requires special wiring rules to be adopted; the tremendous speed is expensive, consequently this type of logic circuit tends to be limited to main frame computers.

In the majority of applications very high speed is not too important and so TTL or CMOS logic gates are very widely and frequently used.

Logic levels

These are the voltages that represent the binary levels '0' and '1'. Using conventional logic (positive logic) binary '0' is represented by a low voltage and binary '1' by a higher (more positive) voltage.

The actual voltage levels are related to that voltage used to supply power to the circuit. In TTL circuits the power supply ($+V_{CC}$) is +5 V ±5%, while CMOS systems will work on supply voltages between 3.0 V and 18 V.

In TTL circuits the logic level '0' could be considered as 0 V while logic level '1' would be defined as +5 V. However there are practical factors which must be considered. To expect voltage levels to be exact is unrealistic, zero volts may actually be 'almost zero', i.e. 0.2 V, while +5 V may in reality be 4.6 V. If the equipment is designed to switch on the specific levels of 0 V and +5 V, operational problems will be caused. Noise will often be present in most systems and this may appear on the pulse waveform as shown in Fig. 2.2. If the equipment 'sees' everything above 0 V as '1' and every-

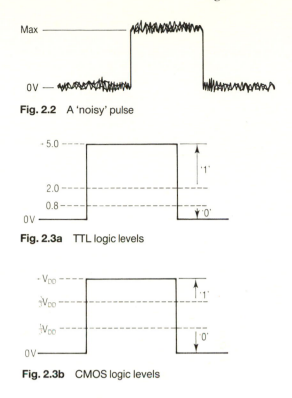

Fig. 2.2 A 'noisy' pulse

Fig. 2.3a TTL logic levels

Fig. 2.3b CMOS logic levels

thing below +5 V as '0' the noise present on a single pulse may well appear as many pulses.

This problem is solved by allocating voltage bands to the logic levels, e.g. for TTL circuits a voltage between 0 V and 0.8 V represents '0', while a voltage above 2.0 V represents '1' as shown in Fig. 2.3(a). CMOS circuits will operate on a supply voltage (V_{DD}) between 3 V and 18 V, consequently the logic thresholds are determined by the actual supply voltage that is used. '0' is represented by voltages between 0 V and 1/3 of V_{DD}, and '1' by voltages between 2/3 of V_{DD} and V_{DD} as shown in Fig. 2.3(b). By allocating voltage thresholds in this fashion very definite regions of the pulse relate to the two logic levels.

THOUGHT _____

So if a +9 V (V_{DD}) supply is used for CMOS circuits '0' will be 0 V to +3 V while '1' will be: +6 V to +9 V? This is true, but in practice it is always sensible to make '0' as close as possible to 0 V and '1' close to 9.0 V, this will ensure reliable operation.

It is possible to obtain most gates in either TTL or CMOS form, but for the sake of simplicity the majority of practical logic investigations in this book will use a power supply of 5.0 V, this means that when a type of gate is specified its TTL or CMOS equivalent can be used with the same power supply.

To help determine the differences between the logic families it is necessary to define the most important characteristics.

Propagation delay

Propagation means 'growth movement' so this term refers to the time that a signal takes to travel through a gate or circuit. To be more precise, a gate cannot switch in zero time, consequently there will be a delay before a logic pulse applied to the input of a gate results in the appearance of a corresponding output pulse. This is illustrated (using an inverter) in Fig. 2.4. Note that the propagation delay time (tpd) is measured from the mid-point (50%) of the waveform. You may notice that the diagram shows two propagation delays, tpd1 is the time that the output takes to change from Logic 1 to Logic 0 while tpd2 is the time that the output takes to change from 0 to 1. Ideally these two times will be the same but in practice they are different. For this reason the propagation delay specified by the manufacturer is the average of these two time periods.

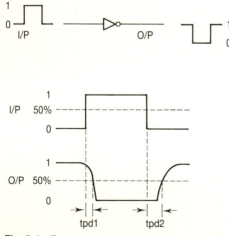

Fig. 2.4 Propagation delay

The propagation delay of modern logic gates is very short (ns) and is quite difficult to measure. Practical Investigation 5 uses discrete components to illustrate this phenomena.

Noise Margin

Some noise is likely to be present on even the best pulse, as illustrated in Fig. 2.2. One reason for using voltage bands for '0' and '1' is to minimize the risk of noise triggering the logic gate. The ability of a logic gate to withstand the presence of noise without triggering or changing its output state is a measure of the gate's 'noise immunity' or 'noise margin' as it is generally known.

THOUGHT _____

So noise margin refers to the maximum noise that can appear at the input of a gate without the output changing? Correct!

_____

This is directly related to the gate logic levels. A TTL gate '0' is represented by voltages between 0 V and +0.8 V while '1' is represented by voltages between +2.0 V and +5 V (Fig. 2.3(a)). This means that the highest permitted voltage for '0' is 0.8 V and the lowest permitted voltage for '1' is 2.0 V. These values are the extremes however and any noise present may certainly alter the gate's operation. Consider the circuit shown in Fig. 2.5.

The maximum '0' voltage for both gates is 0.8 V. The minimum '1' voltage for both gates is 2.0 V. However the output of the first gate is the input of the second. The noise margin refers to the difference between the minimum voltage values for a logic 1 output and input signal, and the maximum voltage values for a logic 0 output and input signal. Study the diagram of Fig. 2.6.

These voltage levels show that the output '1' voltage level can fall to 2.4 V and the output '0'

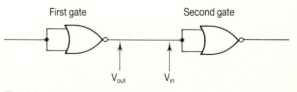

Fig. 2.5 Noise margin voltages

PRACTICAL INVESTIGATION *5*

Propagation Delay

Equipment
Dual beam CRO
Square wave generator
$2 \times$ BC 108 transistors
4×10 k resistors
Power supply
Breadboard

Method
1   Build the circuit shown above.
2   Monitor and record V_{in} and V_{out} with the CRO.
3   Connect the signal generator and adjust it to deliver a 1 kHz square wave of 12 V amplitude.
4   Examine the waveforms noting particularly their time relationships.
5   Increase the input frequency to 10 kHz, sketch the waveforms and measure the propagation delay that occurs (you may have to expand the CRO timebase ($\times 10$) to do this!).
6   Repeat step 5 for input frequencies of 20 kHz and 40 kHz.

Results
1   Is the propagation delay affected by the input frequency?
2   List some effects which propagation delay can cause in switching circuits.

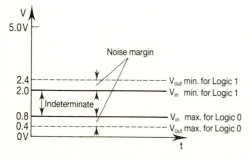

Fig. 2.6   TTL noise margin voltage levels

level can rise to 0.4 V, *but* the input voltage for level '1' can fall to 2.0 V while the input for level '0' can be as high as 0.8 V. The noise margin is given by:

$$V\text{o/p} - V\text{i/p} = 2.4 \text{ V} - 2.0 \text{ V} = 0.4 \text{ V}$$
or
$$0.8 \text{ V} - 0.4 \text{ V} = 0.4 \text{ V}$$

So for TTL gates the noise margin or noise immunity is taken as 0.4 V (400 mV). However please bear in mind that this is the worst possible case; in practice the noise margin of TTL gates is in the region of 1.0 V, i.e. much better than the quoted 0.4 V.

THOUGHT

This is fine because the TTL logic levels are absolute since the power supply is a specified +5.0 V. What about CMOS where the power supply can be any value from +3 V to +18 V? Fig. 2.3(b) showed that the voltage levels for CMOS are such that '0' is less than 1/3 V_{DD} and '1' more than 2/3 V_{DD}. Fig. 2.7 shows that the noise margin for CMOS is thus 1/3 V_{DD}.

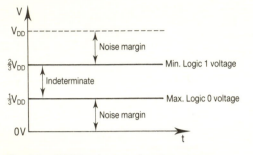

Fig. 2.7   CMOS noise margin voltage levels

THOUGHT

There is a voltage band for both logic families that is labelled 'indeterminate' — what does this mean? This is a voltage range which is neither one level nor the other. The logic levels are clearly defined for all logic families but if a gate was presented with an input voltage in the indeterminate range it could interpret it as a '0' or '1'. The golden rule is to avoid this region by ensuring that inputs to a logic gate or circuit are as close to the specified '0' and '1' levels as possible.

Fan-out

A logic gate is an electronic circuit and as such can only supply a certain amount of output current before its output voltage will be affected. A gate has only one output and this may be connected to the input of a number of other gates (see Fig. 2.8). These gates will each draw current, so there will eventually come a point when the gate to which they are connected will become overloaded and its output may change to an indeterminate level.

The 'fan-out' of a logic gate refers to the maximum number of other similar gate inputs that can be connected to a single gate output without changing its specified logic output level. Typical fan-outs are: TTL = 10 CMOS = 50.

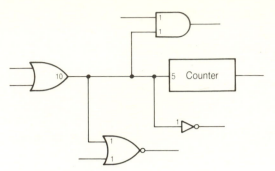

Fig. 2.8   Fan-in and fan-out

Fan-in

One input to a gate is considered to be a standard load and represents a fan-in of 1. A complete logic circuit or network may have only a single input but this will draw more current than a single gate, so it is expressed in terms of the number of standard loads it represents. Thus the counter circuit of Fig. 2.8 draws as much current as five single gate inputs and so it has a fan-in of 5. You can see from Fig. 2.8 that the OR gate has a fan-out of 10 and is supplying a Fan-in of 8.

From the designers point of view it is the 'fan-out' capability of a gate that is of major importance.

Power Dissipation

This refers to the amount of power that a single gate will draw from its power supply. This power demand will be determined by the operating current requirements of the gate, although the power requirement for a single gate may be low it should be remembered that a logic circuit often contains many gates, so the power dissipation may ultimately be quite high.

Before taking a closer look at the gates themselves let us compare the logic families discussed so far in terms of their characteristics.

A cursory glance at Table 2.1 shows that ECL gates are the fastest (i.e. they have the shortest propagation delay time) and this is their chief advantage. In the case of the two readily available types TTL and CMOS, TTL is faster but has a low fan-out and high power consumption; CMOS operates at a third of the speed of TTL

Table 2.1 Logic families compared

Family type	Propagation delay (ns)	Noise margin (V)	Fan-out	Supply voltage	Power dissipation (mW)
TTL	9	0.4	10	5 V	40
CMOS	30	1.6	50	5 V	0.001
ECL	1.1	0.4	50	−5.2 V	30

but offers greater noise immunity coupled with a very much lower power consumption (ideal for battery operation).

THOUGHT

If CMOS can work at higher voltages than TTL (3 V–18 V) do its characteristics remain unchanged? No! — the speed and noise margin of CMOS circuits depend upon the power supply voltage with the fastest operation achieved using a supply in the range of 9.0 V–13 V.

Inside the Gates

To use the logic gates successfully it is not essential to understand the circuitry inside the gate itself. However when designing digital circuits an appreciation of the differences between the families is essential since it helps to predict the operational suitabilities of each type.

The Standard TTL Gate

Silicon bipolar transistors are used with the circuit of Fig. 2.9 showing a standard TTL 3 input NAND gate.

Circuit Operation

Recalling your knowledge of transistor operation; when a current flows in the base of a transistor, collector and emitter currents flow.

Assume that all the inputs are at '1' (A = B = C = +5 V). This effectively reverse biases the base-emitter junction of T1. The base-collector junction is forward biased however so current will flow into the base of T2 turning it on. The collector voltage of T2 falls to almost zero (0.1 V), therefore T3 is off, but some of the emitter current of T2 flows into the base of T4 turning it on, consequently the output Q will be almost zero!

If any of the inputs is at '0' (0 V) transistor T1 will now be forward biased so it will conduct, making the base of T2 low thus turning it off. With T2 off its collector will be higher thus turning T3 on. The output is taken from the emitter of T3 (emitter follower mode) and so the output Q will be high, e.g. '1'.

The totem pole

An intriguing aspect of this NAND gate circuit is the arrangement of the output stage T3 and T4, this is explored further using Fig. 2.10. Imagine that the base of T3 is held at 0 V while the base of T4 is at a higher positive voltage. T3

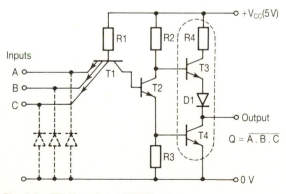

Fig. 2.9 TTL three input NAND gate

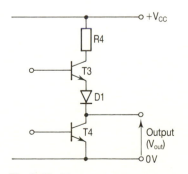

Fig. 2.10 Totem pole arrangement

will be off and T4 will be on. The output will thus be at almost zero volts. If the situation is now reversed and T4 base is held at 0 V while T3 is held high, T3 will conduct but T4 will have a very high resistance; the output will now be high at approximately 4.2 V. Given by:

$$V_{out} = VCC - (0.2\,V + 0.6\,V) =$$
$$5\,V - 0.8\,V = 4.2\,V.$$

THOUGHT

In Fig. 2.9 there are diodes shown connected across the inputs. What is their purpose? These prevent a negative voltage from being applied to the transistor T1 where it would cause damage. They also help to improve the noise margin of the gate.

The CMOS Gate

The letters CMOS stand for 'complementary metal oxide semiconductor'. The transistors will be P and N channel MOSFETS (metal oxide semiconductor field effect transistors). Fig. 2.11 shows the circuit of a 2 input CMOS NOR gate. Inputs A and B are applied to the gate of the respective MOSFET, where the input resistance will be almost infinity. When both inputs are at 0 V (logic 0) the N channel devices T3 and T4 are off while the P channel transistors T1 and T2 are on. The output Q will thus be at approximately $+V_{DD}$, i.e. '1'.

If either input A or B is at '1' ($+V_{DD}$) T3 or T4 will be on while T1 or T2 is off giving a low output ('0'). When both inputs A and B are at '1' ($+V_{DD}$) T1 and T2 will be off while T3 and T4

will be on. Consequently the output Q will be low ('0').

The high input impedance of CMOS gates together with the low power consumption and improved noise margin make them very attractive to the designer. There can however be some problems related to the high input impedance that must be considered.

1   Unused gate inputs can charge to any voltage levels due to pick up or static. Make sure that *all* unused gate inputs are connected either to the 0 V line or $+V_{DD}$ to protect them, i.e. do not leave inputs to a gate floating.
2   Static electricity is always present with potentials of many thousands of volts possible. Many early CMOS gates were inadvertently destroyed by this static potential but today most devices are internally protected (using zener diodes). Even so it makes sense to observe the following rules – just to be on the safe side.
   (a)   Keep CMOS devices in the protective foam or conducting plastic packing that keep the pins short circuited until required.
   (b)   Insert the device last when soldering or building a circuit.
   (c)   During circuit construction earth the soldering iron and the operator for maximum protection from static.

THOUGHT

How do you know if a device is internally protected? CMOS logic gates are available as the '4000' series. There is now available a 4000 B series, the suffix 'B' indicates that the device is buffered and is also internally protected.

The ECL Gate

The name 'emitter coupled logic' tells us that bipolar transistors are used in the construction as with TTL. TTL gates operate by driving the individual transistors into saturation, this means that a delay is introduced during switching while any accumulated charge dies away. ECL circuits use non-saturating transistors thus removing any switching delay. The basic circuit for a two input ECL OR/NOR gate is shown in Fig. 2.12.

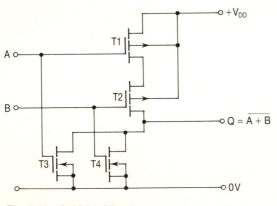

Fig. 2.11   CMOS NOR gate

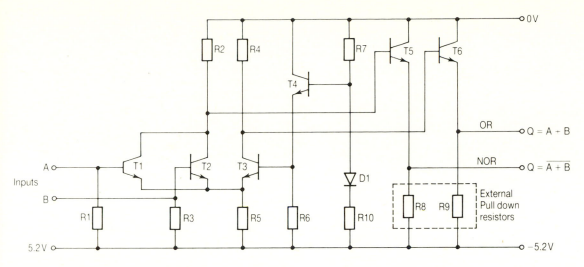

Fig. 2.12 Two input ECL OR/NOR gate

Operation

R7, D1 and R10 hold the base of T4 at a potential of -0.6 V. This in turn keeps the base of T3 at a -1.2 V reference. The diode D1 acts as a temperature compensation device that keeps T3 base potential constant despite temperature variations. The logic levels involved are: Logic $1 = -0.8$ V   Logic $0 = -1.6$ V.

Note This is positive logic! The levels are 0.4 V above and below the -1.2 V reference. If A and B are both at -1.6 V (Logic 0) then the base of T3 will be more positive at -1.2 V, T3 conducts a little more and its collector current develops a voltage across R4 that makes T6 conduct less, giving about -1.6 V (Logic 0) at its emitter. However, at the same time the base potentials of T1 and T2 are more negative than T3 so the collectors of T1 and T2 are at almost 0 V, (remember the supply rail is not $+5$ V but 0 V). This means that T5 base is high and its emitter voltage is about -0.8 V which is Logic 1.

If either A or B is at Logic 1 (-0.8 V) T1 or T2 will conduct more than T3 making the collector of T1,T2 more negative than that of T3, consequently T5 emitter will be at Logic 0 (-1.6 V) and T6 emitter at Logic 1 (-0.8 V).

Note At no time is any transistor driven into saturation, they simply conduct a little more or

less, this is why the circuitry appears to be more complex than either CMOS or TTL.

THOUGHT

Why is a negative supply used? This is just a convention that is used for ECL. A supply rail (V_{CC}) of 0 V and a ground rail of -5.0 V is the same as a V_{CC} of $+5.0$ V and a ground rail of 0 V, isn't it?

Logic families — A further look at TTL

When comparing the various types of gates it would appear that CMOS offers the best compromise. However TTL (marketed under the 74 series label) has been a commercial standard for some years. Development however does not stop simply because new device types appear on the market. The result of this is a logic family within a logic family. So we now have the 74 series family which offers different versions of TTL gates each with their various advantages. To examine these let us first consider the major disadvantages of the standard TTL gates.

1   A propagation delay which is shorter than CMOS but longer than for ECL.

2   Power consumption — very high when compared to equivalent CMOS gates.

Switching speeds

Practical Investigation 5 allows the propagation delay of a transistor switching circuit to be measured. This peculiarity can be further studied using the diagram and waveforms of a transistor switch shown in Fig. 2.13(a) and 2.13(b). When V_{in} changes from 0 V to +5 V there is a time delay (td) before V_{out} starts to fall. The time taken for the voltage to fall from 90% to 10% of its maximum value is the fall time (tf). When V_{in} changes from +5 V to 0 V there is another time delay (ts) as the charge stored in

the base region decays. The final time delay (tr) is the time taken for V_{out} to rise from 10% to 90% of its maximum value.

You may see from the diagram that the major problem is caused by the charge stored in the base region. To reduce this the transistor must be prevented from saturating. A diode connected between the collector and base, as shown in Fig. 2.14(a), stops excess current entering the base region and thus prevents saturation and hence any storage charge from accumulating.

In practice a Schottky diode (Fig. 2.14(b)) is used because this has a very low forward volt

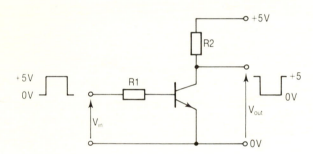

Fig. 2.13a   The transistor as a switch

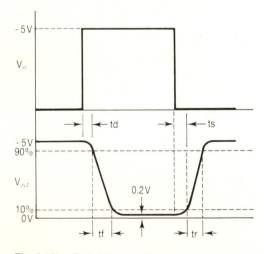

Fig. 2.13b   Switching waveforms

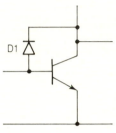

Fig. 2.14a   Anti-saturation diode D1

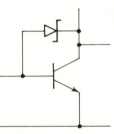

Fig. 2.14b   Schottky diode

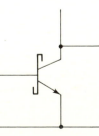

Fig. 2.14c   Schottky-clamped transitor

drop. It is now possible to incorporate this diode into the transistor itself when it is manufactured, so that the device becomes the Schottky-clamped transistor of Fig. 2.14(c) which has a very fast turn on without the turn off delay caused by saturation.

This improvement means that 74 series TTL gates can also be obtained in Schottky form. In addition to this, manufacturing methods have reduced the power consumption to about one fifth of standard TTL. These are marketed under the label of low power Schottky TTL (coded 74LS).

Further improvements on this Schottky circuit have resulted in the Advanced low power Schottky TTL gate (coded 74ALS) which is twice as fast as the 74LS type and uses half the power.

Fast TTL

These circuits use an isoplanar transistor manufacturing process that results in very short propagation delays together with lower power dissipation.

Special TTL Gates

To try and meet the high noise immunity and low power requirements of the CMOS gates, a number of TTL circuits have been made using CMOS techniques. This has given rise to the curious TTL CMOS circuit. Marketed under the label of High Speed CMOS (coded 74HC) this TTL circuit offers the same speed as the LS type TTL but with CMOS compatible inputs. The type coded 74HCT has TTL compatible inputs, and both of these types have outputs that are suitable for either TTL or CMOS applications.

To help clarify things let us use Table 2.2 to compare the various types by looking at the performance characteristics of a Quadruple 2 input NAND Gate. You may see from this that it

is possible to obtain seven different versions of a Quadruple 2 input NAND gate. This wide selection is available for many of the popular logic gates.

Note In addition to the types mentioned so far there is a military range of logic circuits with the prefix 54, e.g. 5400. We shall not be considering these devices.

THOUGHT

How on earth do you make the right choice? In practice it is quite simple; every application has minimum requirements of speed and power consumption so that unless the occasion demands high speed the choice will probably be either standard TTL or CMOS. If TTL is chosen the various TTL families are compatible and are thus interchangeable.

Logic Gates with open collector output

Study the TTL data sheets on page 114 and you will notice that a number of gates are available with open collector outputs, e.g. the 7401 is the same as a 7400 — it is a Quadruple 2 input NAND gate but it has open collector outputs. Likewise 7409 is the same as the 7408 Quadruple 2 input AND gate but with open collector outputs. To illustrate what this means study the circuits of Fig. 2.15(a) and 2.15(b).

The normal TTL NAND gate with a totem pole output circuit is shown in Fig. 2.15(a). While Fig. 2.15(b) shows an open collector TTL NAND gate. You may see that the collector of T3 is left unconnected. To use this gate a pull-up resistor (Rp) must be connected externally. When the collector of T3 is high (Logic 1) the output will be at $+V_{CC}$ because T3 is off, so Rp effectively pulls the output up to the highest voltage which is V_{CC}. If the collector of T3 is low

Table 2.2 Gate characteristics

Type	Code no.	Power consumption	Propagation delay	Supply voltage	Fan-out
CMOS	4011B	1 μW	30 ns	+5 V	50
Standard TTL	7400	40 mW	9 ns	+5 V	10
Low power Schottky	74LS00	2 mW	8 ns	+5 V	10
Advanced low power Schottky	74ALS00	1 mW	4 ns	+5 V	10
High speed CMOS TTL	74HC00	1 μW	8 ns	+5 V	8
High speed CMOS TTL	74HCT00	2 nW	9 ns	+5 V	8
Fast TTL	74F00	5 mW	3.7 ns	+5 V	15

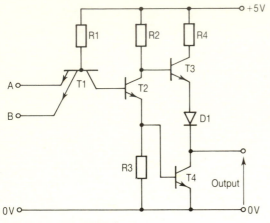

Fig. 2.15a   TTL NAND Gate

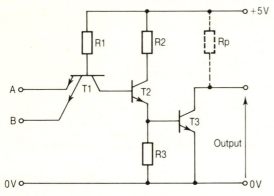

Fig. 2.15b   TTL open collector NAND gate

Suppose a TTL gate was required to supply an input signal into a CMOS gate that had a +12.0 V supply. The logic levels required for the CMOS inputs will be logic 0 = +4 V max. Logic 1 = +8 V minimum. The output from a TTL gate could not operate this CMOS gate. An open collector gate can be used as shown in Fig. 2.16. The external pull-up resistor (Rp) is connected to the CMOS supply rail of +12 V. When the TTL output is Logic 0 the CMOS input is 0 V. When the TTL output is Logic 1 the CMOS input is +12 V — problem solved!

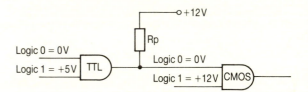

Fig. 2.16   Logic level change using an open collector gate

Paralleled outputs

This configuration is shown in Fig. 2.17 where a pair of two input NAND gates have their outputs connected or 'wired' together. If both Gate outputs are high (Logic 1) the output Q will be high. If either gate output is Low (Logic 0) the output Q will be Low (Logic 0). This gives an overall AND function.

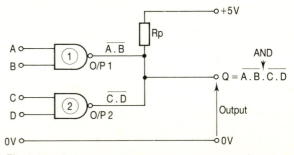

Fig. 2.17   Open collector NAND gate in the wired AND mode

Note   The name of this circuit is determined by the type of logic used.

(Logic 0) the output will be almost zero. (This is because if T3 is on its $V_{CE} \simeq 0.1$ V.)

There are two main applications of this device.

1   To enable a TTL gate to provide an output other than 5.0 V. This is particularly useful when interfacing TTL to devices or circuits. For example, it is often required to connect a number of gates to a data bus. If each gate is connected via a pull-up resistor then at any instant all the gates (except the one being used) present a high impedance (open circuit) to the data bus.

2   To allow the outputs of TTL gates to be connected together thus performing a different logic function.

Positive Logic (Conventional Logic)

Uses the more positive of the voltages to represent Logic 1 and the less positive voltage to represent Logic 0 thus: 0 V = Logic 0 +5 V = Logic 1.

Negative Logic

Uses the reverse of this, thus: 0 V = Logic 1 +5 V = Logic 0.

This means that if the circuit of Fig. 2.17 is operated using positive logic it is called a wired AND operation, since output Q will only be high (Logic 1) if the outputs of both gates are high. If negative logic is used (where 0 V = Logic 1) you may see that if the output of either gate is 0 V (Logic 1) the output Q will be 0 V (Logic 1) so the circuit is called a wired OR operation.

THOUGHT

What should the value of the pull-up resistor be? If Rp is a low value it gives increased speed and noise immunity but places a greater power demand on the gate. If Rp is a large value this results in a slow operating speed, poor rising edge, and a lower noise margin. Typical values are between 470R and 5k0.

Can you tell if a gate has an open collector output from the symbol? A half circle drawn on the symbol or an asterisk drawn as shown in Fig. 2.18 indicates an open collector arrangement, however these representations are by no means universal with the only true indication given by the gate code number and data sheet.

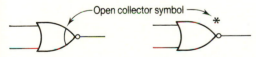

Fig. 2.18   Open collector representation

Self Assessment 2

1   Write the logic output for the wired gate circuit shown in Fig. 2.19.
2   Name the type of wired gate if:
(a)   positive logic is used;
(b)   negative logic is used.

Practical Investigation 6 enables you to study these wired gates.

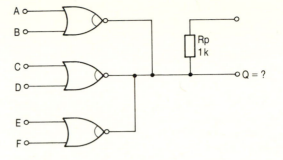

Fig. 2.19

WARNING!

Do not attempt to parallel the outputs of conventional TTL or CMOS gates since they have not got open circuit outputs. The result would be excessive current flow in the gate output and the demise of the gate!

Although the paralleling of gates enables complex functions to be constructed cheaply, there is a major disadvantage. If one gate becomes faulty it is very difficult to perform fault diagnosis on the circuit, since all the gates will be held low by the 'rogue' gate.

Tristate Logic

Open collector logic requires the use of pull-up resistors, this tends to degrade the speed of operation and the noise immunity. A solution to this problem is to use Tristate logic gates. This term is very misleading because it implies that three logic levels are used. This is not so! Tristate gates have the same function as normal gates except that the output can be switched into a high impedance state using the control input. Fig. 2.20 shows tristate versions of the OR and AND gate. If the control input is at Logic 1

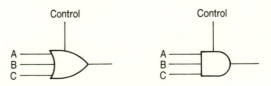

Fig. 2.20   Tristate logic gates

PRACTICAL INVESTIGATION 6

Wired AND Gates

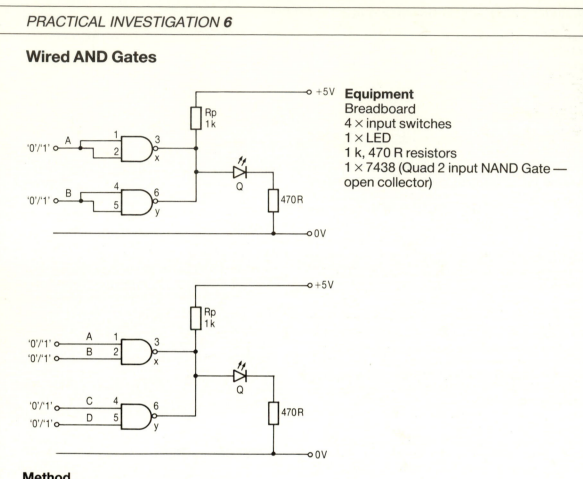

Equipment
Breadboard
4 × input switches
1 × LED
1 k, 470 R resistors
1 × 7438 (Quad 2 input NAND Gate — open collector)

Method
1   Construct circuit A shown above.
2   By investigation complete the truth table shown below:

A	B	Q
0	0	
0	1	
1	0	
1	1	

3   Construct circuit B.
4   Draw up, and by investigation complete, the full truth table for this logic circuit (four input variables means 16 different possible input variations).

Results
1   What gate function is performed by circuit A?
2   Study these two circuits closely and note that regardless of the input conditions, for the output Q to be '1' both gate outputs must be '1', $\therefore$ Q = x . y.

they behave normally. When the control input is taken to Logic 0 the gate outputs assume a high impedance state.

You will find that a number of TTL devices are available with tristate outputs and these are often used in data selection and highway based systems.

Assertion Level Logic Notation

There is a peculiarity concerning logic levels that is worth mentioning at this stage since it will crop up periodically and may cause confusion.

Positive logic means that:
the higher (more positive) voltage level = Logic 1;
the lower (more negative) voltage level = Logic 0.
Negative logic is the reverse of this convention so:
the higher (more positive) voltage level = Logic 0;
the lower (more negative) voltage level = Logic 1.

A gate or device that operates from a logic 1 can be described as being active high, while a logic 0 triggered gate or device is said to be active low. The standard logic symbol assumes positive or active high logic. It is acceptable to show logic gates that are active low with a negating circle drawn on the inputs as shown in Fig. 2.21. This is called assertion level logic. Look out for the active low symbol on gates and devices in the coming chapters and data sheets.

Active low buffer   Active low NOR gate

Fig. 2.21   Assertion level logic

Logic Gate Review

1   A gate is an integrated circuit that can have a number of binary inputs but only one binary output.

2   American Standard Symbols (MIL/ANSI) are used almost universally for logic gates.

3   Logic families consist of TTL, ECL, and CMOS types.

4   There are two logic levels: Low ('0') and High ('1').

5   Logic levels are represented by voltage bands:
Logic 0 = from 0 V to +0.8 V   Logic 1 = from +2.0 V to +5.0 V for TTL.
Logic 0 = from 0 V to $1/3\ V_{DD}$ Logic 1 = from $2/3\ V_{DD}$ to V_{DD} for CMOS.

6   Negative logic levels are the inverse of those for positive logic. With negative logic the more negative level = Logic 1 while the less negative level = Logic 0.

7   Propagation delay is the time delay between a logic pulse being applied to the gate input and the corresponding pulse appearing at the output.

8   Noise margin or noise immunity refers to the maximum noise signal that can appear at the gate input without the output changing. The higher the noise immunity the better.

9   The quoted noise margin of TTL gates is 0.4 V, for CMOS it is $1/3\ V_{DD}$ — in practice it is usually much better (higher) than this value.

10   Fan-in is the loading that a gate will present to any circuit supplying an input signal. A TTL inverter has a fan-in of 1, i.e. 1 input = a fan-in of 1. A four input NAND gate has a fan-in of 4.

11   Fan-out is the load that a gate can supply without its output state changing, i.e. a TTL gate with a fan-out of ten can drive the inputs of ten other TTL gates and yet operate correctly.

12   Power dissipation is the amount of power a gate will draw from its d.c. power supply. CMOS offers the lowest power dissipation of about 1 μW per gate.

13   Within the TTL logic families there area number of different types of gate performing the same function as standard TTL but offering various other advantages.
Low power Schottky TTL coded LS, e.g. 74LS00, offering faster speed and requiring half the power of standard TTL.
Advanced low power TTL coded ALS, e.g. 74ALS00.

Fast TTL coded F, e.g. 74F00.
These types offer increased speed and reduced power consumption.

14   There are two types of TTL made specific-
ally to be compatible with CMOS.
TTL HC, e.g. 74HC00, CMOS compatible
inputs.
TTL HCT, e.g. 74HCT00, TTL compatible
inputs.
Both have outputs suitable for driving TTL
or CMOS gates.

15   TTL gates are available with open collector
outputs that require a pull-up resistor to be
externally connected. This allows:
  (a) Gate outputs to be paralleled or wired
       together to perform a specific logic
       function.
  (b) A gate's output logic level to be
       changed.

16   If a number of NAND or NOR gates have
their outputs paralleled together the result
is a *wired* AND gate for positive logic, or
alternatively a wired OR gate for negative
logic.

17   External pull-up resistors can only be used
with open collector gates. Standard TTL
and CMOS gates will be destroyed if wired
in this manner.

18   Tristate logic devices are obtainable that
perform exactly like normal gates, but have
a control input that enables their outputs to
be switched into a high impedance state.

19   Assertion level logic can be used to show
when a device operates or is switched using
the more negative logic level (active low).

Self Assessment Answers

Self Assessment 2

1

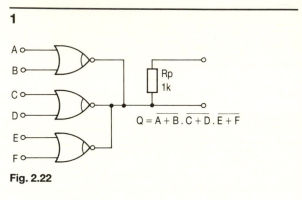

$$Q = \overline{A + B} . \overline{C + D} . \overline{E + F}$$

Fig. 2.22

2   The output (Q) will be logic 1 only when all the
gate outputs are at logic 1. It is thus a wired
AND gate for positive logic. Alternatively if
negative logic is used the output will be a low
voltage (Logic 1) when any gate output is a
low voltage (Logic 1) so for negative logic it is
a wired OR gate.

3
Combinational Logic Networks

'Logic gates can be interconnected or combined to give any desired logic function.'

It has been shown that the function of a gate can be explained using a truth table and a Boolean expression, indeed the Boolean expression is often derived from a truth table. Consider the logic circuit of Fig. 3.1(a). This is a logic circuit with three input variables or literals. A truth table can be drawn up for all possible input conditions, as shown in Fig. 3.1(b).

Note that each binary variable has two possible values and as there are three input variables the total number of possible input variations is given by $2^3 = 8$. Notice also the way the table is drawn up: column C changes state with each row, column B changes every 2 rows while column A changes every four rows. By drawing the table in this way every one of the eight input variations is used in a binary sequence representing denary numbers from 0 to 7.

From the truth table a Boolean expression for the output of the circuit can be derived:

$$Q = (\bar{A}.\bar{B}.C) + (\bar{A}.B.C) + (A.B.C)$$

Notice how brackets (parentheses) are used to group the product terms together and separate them. This type of Boolean expression is known as a 'sum of products' expression. An expression for the output can also be found by writing the Boolean expression at the input and the output of each gate and building up the total expression as shown in Fig. 3.2. So now we have a Boolean expression derived from the truth table and one by inspection of the logic diagram. They are not the same! Or are they?

The expression $Q = (\bar{A} + B).C$ is a simpler one than

$$Q = (\bar{A}.\bar{B}.C) + (\bar{A}.B.C) + (A.B.C)$$

and since it refers to the same logic circuit it must be a simplified or minimized version of the

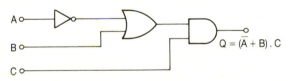

Fig. 3.1a   Truth table circuit

Input variables			Out-put	
A	B	C	Q	
0	0	0	0	
0	0	1	1	$\bar{A}.\bar{B}.C$
0	1	0	0	
0	1	1	1	$\bar{A}.B.C$
1	0	0	0	
1	0	1	0	
1	1	0	0	
0	1	1	1	A.B.C

Fig. 3.1b   Truth table

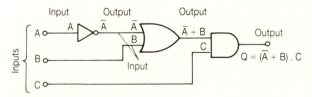

Fig. 3.2   Deriving the Boolean output

expression derived from the truth table. Minimization is important in logic circuit design and will be fully explained later; but for the moment let us apply the techniques discussed so far to solve a specific logic problem.

A basic burglar alarm circuit consists of the following arrangements:

1   An infra-red detector device produces a normally low output. If intruders encroach within 5 metres of the device they are detected and the output changes to a high.
2   All the doors and windows are fitted with a closed ring circuit, i.e. the loop is continuous, producing a high output. If a door or window is opened the output falls to a low, since the closed ring circuit is now opened.
3   Inside the house there is an acoustic device that normally provides a low output. If a window is broken the sound is detected and the output is high.

The output of each of these detecting systems is fed to a control box, if any security is breached the control sounds the alarm. To set the alarm system there is a keyswitch which supplies the circuit with power when it is in the enabled or armed position.

Requirement
1   Draw up a truth table for the system and from this derive the Boolean expression.
2   Draw the logic circuit for the control unit.

Procedures
1   Establish the rules of the game by naming the inputs and the levels that the logic system will require.
    Let low = Logic 0 and High = Logic 1. This is positive logic!
    There will be one input to the control unit from each of the three detecting systems.
    So let the infra red detector output be A, the window and door ring circuit output be B and the acoustic detector output be C.
    The control activates the alarm, which we can call Q, so that when Q = '0' the alarm is off, when Q = '1' the alarm is on. There is the keyswitch input to the control that enables (arms) or disables the alarm, we shall call this D.
    When the power is off: D = '0' (system disabled); and when the power is on: D = '1' (system enabled).
2   The truth table for the system can now be drawn up as shown in Fig. 3.3.

Inputs			Output	
A	B	C	Q	
0	0	0	1	$\bar{A}\bar{B}\bar{C}$
0	0	1	1	$\bar{A}\bar{B}C$
0	1	0	0	
0	1	1	1	$\bar{A}BC$
1	0	0	1	$A\bar{B}\bar{C}$
1	0	1	1	$A\bar{B}C$
1	1	0	1	$AB\bar{C}$
1	1	1	1	ABC

Fig. 3.3   Security system truth table

Note   The system will not work at all unless the power is on, therefore we shall only need to examine the requirements and conditions for inputs A, B, C.

3   From the details given the alarm will be activated whenever input A is '1' or input B = '0' or input C is '1'. So Q can be entered accordingly and the Boolean expression derived giving:

$$Q = (\bar{A}.\bar{B}.\bar{C}) + (\bar{A}.\bar{B}.C) + (\bar{A}.B.C) + (A.\bar{B}.\bar{C}) + (A.\bar{B}.C) + (A.B.\bar{C}) + (A.B.C)$$

This is a rather long expression that requires some clarification.

4   When presented with a situation always attempt to simplify using common sense, i.e. look at the single output when Q = '0' it is just as valid to say $\bar{Q} = \bar{A}.B.\bar{C}$ or, from section 3, the alarm will sound whenever A = '1' or B = '0' or C = '1'.
    Therefore Q = A + $\bar{B}$ + C.

THOUGHT _____

This looks very similar to $\bar{Q} = \bar{A}.B.\bar{C}$. Yes it does; in fact $\bar{Q} = \bar{A}.B.\bar{C}$ describes the alarm off state while Q = A + $\bar{B}$ + C describes the alarm on state.

5   A logic circuit to fulfil this function is shown in Fig. 3.4.

The complete control circuit consists of a three input OR gate and an inverter. Remember the keyswitch D merely turns on the power supply to arm the circuit.

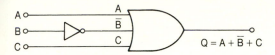

Fig. 3.4   Security alarm circuit

This rather long-winded exercise serves to illustrate a number of things.

1   Verbal or written conditions can be represented by a truth table.
2   The Boolean expression for the output of a circuit may be very long but the expression for the inverse of the output may be short.
3   It is possible to obtain a Boolean expression without a truth table provided the circuit requirements are clearly established.
4   Boolean expressions that appear to be totally different may mean the same thing.

THOUGHT

Is there any way of telling when totally different expressions mean the same thing? Yes, but only when you are aware of the techniques used to minimize Boolean expressions.

Minimization Methods

There are two main methods of reducing Boolean expressions to their simplest forms, one involves the already mentioned Boolean algebra and the other uses Mapping techniques. It is very important to realize that care taken to develop minimization skills at this stage will be amply rewarded later on by the ease with which it will be possible to interpret and design logic circuitry.

Boolean Algebra

This presents minimal problems for those skilled in normal algebra. However, many people on hearing the word algebra mentioned 'come over faint' and shy away. To these individuals I say "Bear with me, this is not as bad as you think"! Let us start by stating the obvious. In Boolean expressions the plus (+) sign means OR the dot (.) means AND. So $A+B$ means A

OR B; $A.B$ means A AND B. The 'dot' is the symbol for multiplication, if preferred this can be assumed and not written in a Boolean expression. Therefore $(A.B.C)+(A.D)$ can be written as $(ABC)+(AD)$. Inversion or negation is indicated by a bar across the top of the inverted term or expression, e.g. $\bar{A}$ or $\overline{A+B}$. Therefore $'\bar{1}' = '0'$ and $'\bar{0}' = '1'$. A double inversion cancels out $\bar{\bar{A}} = A$ or $\overline{\overline{A+B}} = A+B$.

Remember Logical 0 ('0') is not the same as mathematical 0 (zero) similarly Logical 1 ('1') is not the same as mathematical 1.

The rules of Boolean algebra are expressed by the Boolean or Logic identities, see Table 3.1.

Table 3.1   Some Boolean Identities

The AND function	The OR function
$'1'.'1' = '1'$	$'1'+'1' = '1'$
$'1'.'0' = '0'$	$'1'+'0' = '1'$
$'0'.'1' = '0'$	$'0'+'1' = '1'$
$'0'.'0' = '0'$	$'0'+'0' = '0'$
$A.A = A$	$A+A = A$
$A.'1' = A$	$A+'0' = A$
$A.'0' = '0'$	$A+'1' = '1'$
$A.\bar{A} = '0'$	$A+\bar{A} = '1'$
$A.B.C = C(A.B) = A(B.C)$	$A+B+C$
	$= (A+B)+C = A+(B+C)$
$A.B = B.A$	$A+B = B+A$
Combinations of functions	*DeMorgans Rules*
$A.B+A.C = A(B+C)$	$\overline{A+B} = \bar{A}.\bar{B}$
$A+A.B = A$	$\overline{A.B} = \bar{A}+\bar{B}$
$A+B.C = (A+B).(A+C)$	
$A+AB = A+B$	
$A(B+C) = AB+AC$	

These rules allow you to manipulate expressions by showing what is possible. For example if you have an expression $Q = A+(A.B)$, this will minimize to $Q = A$ — nothing magical about that!

Suppose a lamp will light if the input conditions are A OR, A AND B then clearly, provided input A is present, the lamp will light regardless of the state of input B. This can be shown as:

$$A+(A.1) = A+1 = A; A+(A.0) = A+0 = A$$

DeMorgans laws are most important in that they allow us to indulge in conversion from one type of gate to another, so that ultimately any circuit can be built from one gate type only (usually NAND or NOR gates).

Now that the logic identities have been introduced, let us use them to manipulate some logic expressions. A good example is the 'exclusive OR' function that is defined by the Boolean expression $A \oplus B$. You may remember I said on page 14 that this is not a true Boolean expression because it cannot be used algebraically. The truth table for the exclusive OR gate is shown in Fig. 3.5. The true Boolean expression for the output is given by $Q = (\bar{A}.B) + (A.\bar{B})$.

A	B	Q	
0	0	0	
0	1	1	$\bar{A}.B$
1	0	1	$A.\bar{B}$
1	1	0	

Fig. 3.5   Exclusive OR truth table

An alternative expression for this gate is $Q = \overline{(AB)} + \overline{(\bar{A}\bar{B})}$. Using the logic identities it should be possible to prove that $Q = (\bar{A}B) + (A\bar{B}) = \overline{(AB)} + \overline{(\bar{A}\bar{B})}$.

1 Take the expression $Q = \overline{(AB) + (\bar{A}\bar{B})}$ and use DeMorgans theorems:
   Break the bar and change the sign to give:
   $$Q = \overline{(AB)} . \overline{(\bar{A}\bar{B})}$$

2 Use DeMorgans again to give:
   $$Q = (\bar{A} + \bar{B}) . (\bar{\bar{A}} + \bar{\bar{B}})$$
   Note the double inversion which cancels to give:
   $$Q = (\bar{A} + \bar{B}) . (A + B).$$

3 We now have two OR functions linked by an AND
   $$Q = (\bar{A} + \bar{B}) . (A + B).$$
   Note This is known as a PRODUCT OF SUMS expression.

4 Multiply out the brackets:
   $$Q = \bar{A}A + \bar{A}B + \bar{B}A + \bar{B}B$$

5 There are now two terms that equal '0' $\bar{A}A = '0'$, $\bar{B}B = '0'$. (If in doubt check the list of logic identities.)
   Therefore, $Q = \bar{A}B + \bar{B}A$ and, since $\bar{B}A = A\bar{B}$ we have shown that $Q = \bar{A}.B + A.\bar{B}$ is   the same as $Q = \bar{A}B + A\bar{B}$.

Self Assessment 3

1 Using a truth table derive a true Boolean expression for the exclusive NOR gate $\overline{A \oplus B}$.
2 Show using Boolean algebra that $\overline{A \oplus B} = \overline{\bar{A}B + A\bar{B}}$.

Boolean algebra can be used to reduce very long expressions to their simplest form. Look back at the truth table of Fig. 3.1(b). This yields the Boolean expression

$$Q = (\bar{A}\bar{B}C) + (\bar{A}BC) + (ABC)$$

yet by inspection of the logic circuit a simpler expression was derived showing that $Q = (\bar{A} + B).C$. If we only had the truth table and not the logic circuit it should be possible to arrive at the minimized expression by using Boolean algebra.

$$Q = \bar{A}\bar{B}C + \bar{A}BC + ABC$$
$$Q = \bar{A}C(\bar{B} + B) + BC(\bar{A} + A)$$
$$Q = \bar{A}C + BC$$
$$Q = C(\bar{A} + B)$$

Therefore $Q = (\bar{A} + B).C$.
Caution The purpose of brackets (parentheses) is to remind us that terms are grouped. Once this rule is accepted the brackets can be left out of the written expression on the strict understanding that they are assumed to be there mathematically. Consequently always be on the lookout for Boolean expressions written without brackets.

Self Assessment 4

1 Use Boolean algebra to minimize the expression

$$Q = AB\bar{C} + AB\bar{C} + A\bar{B}C + \bar{A}BC.$$

Boolean algebra is undoubtedly satisfying to perform but it does rely on the intuition, skill and experience of the manipulator, who must know which identities to use and when to indulge in a bit of multiple negating in order to progress with the simplification. What this means in reality is that unless you are expert at Boolean algebra you will have minimal confidence in your minimized answer, however short it may be! Happily for practical and design purposes the mapping method for minimization yields quick and accurate results but you must at all times be aware of DeMorgans rules.

MAPPING TECHNIQUES

A map for logic minimization consists of a grid on to which can be plotted each individual expression of a Boolean equation. By grouping these terms a simplified expression can be arrived at quickly and easily.

The most popular mapping method uses the Karnaugh (pronounced ('Car-no') map. For a two variable Boolean expression the four cell grid shown in Fig. 3.6(a) is required. Each ex-

pression for a Boolean equation can be entered in the appropriate cell as a logic 1, e.g. the expression $Q = (\bar{A}.B) + (\bar{A}.\bar{B})$ can be mapped as shown in Fig. 3.6(b).

These cells can now be grouped by determining the term common to the linked cells, in this case the only common term is $\bar{A}$ so the expression has been minimized from $Q = (\bar{A}.B) + (\bar{A}.\bar{B})$ to $Q = \bar{A}$. To be perfectly correct this type of map was originally called a Veitch diagram. Today, however, it is almost universally recognized as a Karnaugh map.

A Karnaugh map can readily be drawn for up to four variables, although it is possible to simplify up to eight variables by using a combination of maps. However, let us now look at the world of mapping in greater detail.

Consider the three variable input map shown in Fig. 3.7 which has the expression $Q = (AB\bar{C}) + (A\bar{B}\bar{C}) + (\bar{A}B\bar{C}) + (\bar{A}\bar{B}\bar{C})$ entered on it. While this is perfectly adequate it is a bit confusing establishing the identity of each cell. An alternative method of representation labels the cells with their logic levels as shown in Fig. 3.8.

Notice how the adjacent cells are numbered so that in going from one cell to the next only one variable changes state. Using this format a

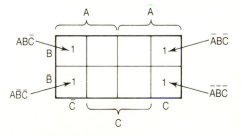

Fig. 3.7 Three variable map

Fig. 3.6a Two variable Karnaugh map

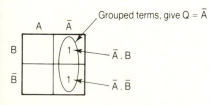

Fig. 3.6b Plot of $Q = \bar{A}.B + \bar{A}.\bar{B}$

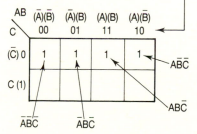

Fig. 3.8 Alternative three variable input map

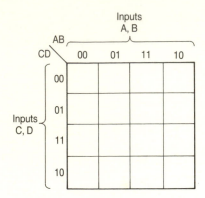

Fig. 3.9a   Four variable input map

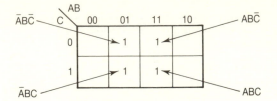

Fig. 3.9c   Plot of Q = ĀBC + ĀBC̄ + ABC̄ + ABC

four variable input map will look like that shown in Fig. 3.9(a).

Before we use this mapping technique there are some simple rules for grouping that must be understood. These are listed below:

1   Grouping of terms must be carried out with regard to their binary weightings (powers of 2), i.e. groups of 2, 4, 8, etc. cells may be linked together.
2   The largest group gives the simplest logic function.
3   The map can be considered as a sphere that is opened out, this means that all the edges connect up. Grouping can take place from top to bottom, side to side and corner to corner.
4   When grouping, a cell may be part of more than one group.

Let us put all this into operation and do some minimizing.

The truth table relating to a combinational logic circuit is shown in Fig. 3.9(b) — from this a Boolean expression is derived giving

$Q = (\bar{A}BC) + (\bar{A}B\bar{C}) + (AB\bar{C}) + (ABC)$. This can be plotted on the Karnaugh map of Fig. 3.9(c) either directly from the truth table or the Boolean expression.

Note   To fulfil the truth table completely all the combinations for Q = '0' should be plotted on the map. I have chosen to enter only the combinations where Q = '1'.

The cells can now be grouped as shown in Fig. 3.9(d) and the common terms identified. So the expression
$Q = (\bar{A}BC) + (\bar{A}B\bar{C}) + (AB\bar{C}) + (ABC)$
minimizes to Q = B.

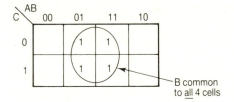

Fig. 3.9d   Grouping to give Q = B

Now let us try the four input expression

$Q = (\bar{A}\bar{B}\bar{C}\bar{D}) + (A\bar{B}\bar{C}D) + (A\bar{B}C\bar{D}) + (\bar{A}\bar{B}C\bar{D})$

Draw the map, plot the terms and then group as indicated in Fig. 3.10. The minimized expression here is $Q = \bar{B}.\bar{D}$.

The maps of Figs 3.11, 3.12 and 3.13 show how four variable complicated expressions can be minimized. I have not shown the original Boolean expressions as they contain many terms. If you wish, you can write these for yourself by deriving the Boolean term for each cell that has a '1' in it, i.e. where Q = '1'.

It is worth remembering that the more groups you have the longer will be the minimized expression, so always attempt to find the largest possible grouping. The greater the number of

A	B	C	Q	
0	0	0	0	
0	0	1	0	
0	1	0	1	ĀBC̄
0	1	1	1	ĀBC
1	0	0	0	
1	0	1	0	
1	1	0	1	ABC̄
1	1	1	1	ABC

Fig. 3.9b   Logic circuit truth table

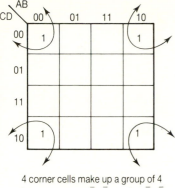

4 corner cells make up a group of 4
Common term = $\bar{B}.\bar{D}$ $Q = \bar{B}.\bar{D}$

Fig. 3.10 Corner grouping

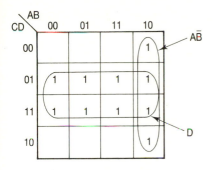

Fig. 3.11 Plot giving $Q = A\bar{B} + D$

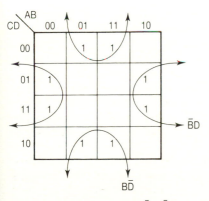

Fig. 3.12 Plot giving $Q = B\bar{D} + \bar{B}D$

'1's linked together, then the greater will be the reduction in variables in that grouping.

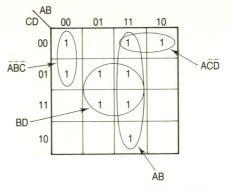

Fig. 3.13 Plot showing $Q = \bar{A}\bar{B}\bar{C} = A\bar{C}\bar{D} + BD + AB$

THOUGHT
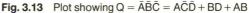

These maps show four variables, can truth tables be drawn for four input systems? Yes they can but a fully specified four input truth table has 16 possible input variations, and that is a large table to draw! An abridged table can sometimes be drawn up however, consider the next example.

A numerically controlled machine tool has four protection sensors that prevent the cutters from being operated in dangerous or damaging situations. The cutters can only be turned on when the output (Q) from the control systems is a logic 1. When the output Q is '0' the machinery will not operate. The safe operating conditions are defined in the truth table of Fig. 3.14.

A	B	C	D	Q
1	0	0	1	1
0	1	1	0	1
1	1	0	0	1
1	0	1	1	1
1	1	1	0	1

Fig. 3.14 Abridged truth table

Under *all* other input conditions the output Q will be '0'. This table can now be plotted on the Karnaugh map, Fig. 3.15, grouped and minimized in the usual way. The minimized expression gives: $Q = A\bar{B}D + AB\bar{D} + BC\bar{D}$.

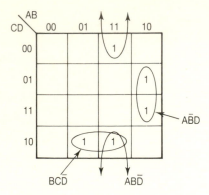

Fig. 3.15   Plot of the truth table in Fig. 3.14

The inverse function

Occasionally a Boolean expression has many input conditions that will give a logic 1 output and only a few that will produce a logic '0' output. In a case like this it is often sensible to minimize the expression in terms of its inverse or complementary function, i.e. instead of minimizing the expression to give Q minimize it to give $\bar{Q}$.

Refer to the truth table of Fig. 3.3 which has all but one input condition giving Q = '1'. It was easier to derive the Boolean expression for the $\bar{Q}$ output than for the Q output.

A Karnaugh map can be used to produce the inverse function. Consider the map of Fig. 3.16. A logic 0 has been entered in all the empty cells. If these cells are now grouped as shown the result is the inverse or $\bar{Q}$ function showing that $\bar{Q} = BD$.

Interestingly the fully specified map of Fig. 3.16 performs according to DeMorgans rules.

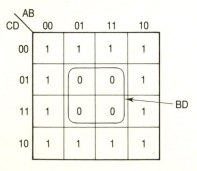

Fig. 3.16   Inverse function map giving $\bar{Q}$ = BD

The inverse function $\bar{Q} = B.D$ means that $Q = \overline{B.D}$, consequently $Q = \bar{B} + \bar{D}$. We can now see that duality exists and that $Q = \overline{B.D} = \bar{B} + \bar{D}$.

Self Assessment 5

1 (a) Using a Karnaugh map minimize the expression
$Q = ABC + \bar{A}B\bar{C} + \bar{A}BC + \bar{A}B\bar{C} + \bar{A}\bar{B}C + A\bar{B}C$

   (b) From the map determine the inverse function of the expression in (a).

2   The truth table of Fig. 3.17 shows the output of a logic circuit. Under all other input conditions Q = '1'. Using a Karnaugh map produce a minimized expression for Q.

A	B	C	D	Q
0	0	0	0	0
1	0	0	0	0
1	1	0	0	0
0	0	1	0	0
1	1	1	0	0
1	0	1	0	0

Fig. 3.17

The 'Don't Care' or 'Can't Happen' Situation

In many logic situations, particularly those involving a large number of inputs, there may be situations where not all the possible input combinations will be used. This means there may be a combination of inputs that will never occur (can't happen) so the output with this combination cannot be specified. Alternatively there may be input combinations that may or may not occur; either way the output state does not matter (don't care). Consider the truth table of Fig. 3.18.

This is the fully specified truth table for a four input combinational logic circuit. The outputs with an X indicate input combinations that either cannot happen, or if they do the state of the output (Q) can be taken as either '1' or '0'. The Karnaugh map can be drawn in the usual way but in this case all the cells should be marked with the appropriate output i.e. 1 or 0 or

A	B	C	D	Q	
0	0	0	0	0	
0	0	0	1	0	
0	0	1	0	1	$\bar{A}\bar{B}C\bar{D}$
0	0	1	1	0	
0	1	0	0	1	$\bar{A}B\bar{C}\bar{D}$
0	1	0	1	1	
0	1	1	0	X	
0	1	1	1	X	
1	0	0	0	0	
1	0	0	1	0	
1	0	1	0	X	
1	0	1	1	X	
1	1	0	0	X	
1	1	0	1	1	$AB\bar{C}D$
1	1	1	0	1	$ABC\bar{D}$
1	1	1	1	X	

Fig. 3.18 Four input truth table with 'don't care' states

X as shown in Fig. 3.19a. Since all the cells marked X indicate 'don't care' situations we can allow these cells to be either '1' or '0' (but not both) in order to create the most suitable grouping for minimization as shown in Fig. 3.19(b).

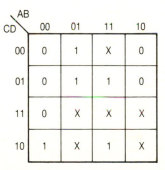

Fig. 3.19a Plot showing 'don't care' terms

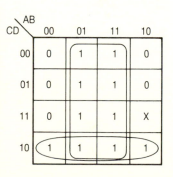

Fig. 3.19b Minimization using 'don't care' terms

The expression can now be reduced to:

$$Q = B + C\bar{D}.$$

It is not necessary to use all Xs only those that are required to help group all the 1s.

Practical Logic Circuit Design

Let us now consider the design of a Logic circuit from the truth table through to the completed combinational circuit. Fig. 3.20 shows the truth table together with the output that the logic circuit must produce.

A	B	C	Q
0	0	0	0
0	0	1	1
0	1	0	1
0	1	1	0
1	0	0	1
1	0	1	1
1	1	0	0
1	1	1	1

Fig. 3.20 Truth table for design assignment

Method
1 Obtain the Boolean expression for the output
   $$Q = \bar{A}\bar{B}C + \bar{A}B\bar{C} + A\bar{B}\bar{C} + A\bar{B}C + ABC.$$
2 Minimize using a Karnaugh map (Fig. 3.21).
3 Design a logic circuit to meet this requirement.

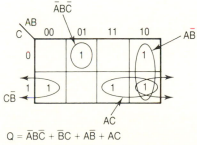

$Q = \bar{A}B\bar{C} + \bar{B}C + A\bar{B} + AC$

Fig. 3.21 Plot of the Boolean expression

The circuit of Fig. 3.22 does fulfil the requirements and you will see that it requires three single inverters, a three input AND gate, three two input AND gates and a four input OR gate. The data sheets on pages 114 and 115 show that these gates are readily available, but it would be

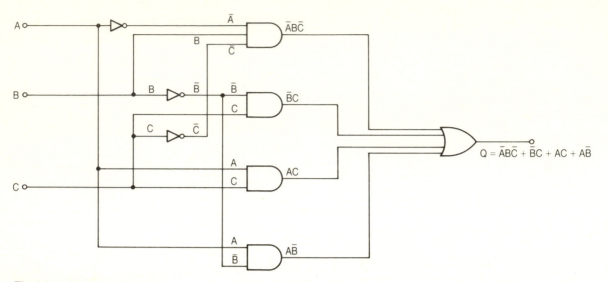

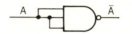

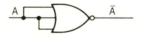

Fig. 3.22   Complete logic circuit

most convenient if this circuit could be designed using only one type of gate.

THOUGHT

Since all sorts of gates can be obtained what is the benefit of designing a circuit so that it uses only one type? Only one type of gate need be stocked in the knowledge that it can be used in any circuit, this is known as 'universal logic'.

Universal Gates

The desire to standardize so that only one type of gate is required is a powerful one. It is possible to design any combinational logic circuit so that it can be built using either NAND gates only or NOR gates only, so these are often referred to as universal gates. To implement a logic function using one type of gate requires the application of DeMorgans rules (Figs 3.23, 3.24 and 3.25). Show how standard gate functions can be implemented using these universal gates.

Although two input gates are shown in the examples, gates with any number of inputs can be used provided all the inputs are paralleled when making an inverter. To convert a logic expression or circuit so that it can be implemented using one type of gate may seem

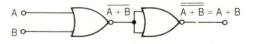

Fig. 3.23a   NAND gate inverter

Fig. 3.23b   NOR gate inverter

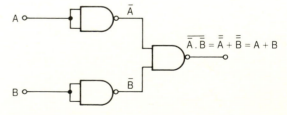

Fig. 3.24a   OR function using NOR gates

Fig. 3.24b   OR function using NAND gates

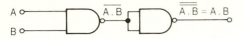

Fig. 3.25a AND function using NAND gates

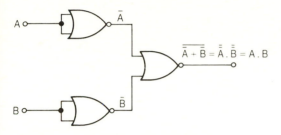

Fig. 3.25b AND function using NOR gates

difficult so it is worth highlighting the salient points.

1 An inverter can be made from a NAND or NOR gate with all the inputs connected together.
2 An OR function can be produced by:
   (a) inverting the output of a NOR gate;
   (b) inverting the inputs to a NAND gate.
3 An AND function can be produced by:
   (a) inverting the output of a NAND gate;
   (b) inverting the inputs to a NOR gate.

With these rules firmly in mind let us now perform some gate implementation exercises in Practical Investigation 7.

THOUGHT _____

To implement the exclusive OR using NAND gates only requires five two input gates, but using NOR gates only it requires six two input gates. Does this mean that the NAND gate implementation is better? In the case yes, but, if you were to implement the exclusive NOR (XNOR) function you would find it would require six two input NAND gates but only five two input NOR gates. So the general rule is:

NAND gates are better when implementing positive functions.
NOR gates are better when implementing negative functions.

Remembering that better simply means fewer gates will be required.

For good measure let us use NAND gates to implement the logic circuit of Fig. 3.22.

1 The logic expression is

$$Q = \bar{A}B\bar{C} + \bar{B}C + AC + A\bar{B}.$$

2 This uses three inverters, one three input AND gate, three two input AND gates and a four input OR gate.

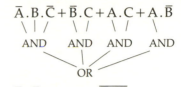

3 $\bar{A}B\bar{C}$ becomes $\overline{\bar{A}B\bar{C}}$
$\bar{B}C$ becomes $\overline{\bar{B}C}$
AC becomes $\overline{AC}$
$A\bar{B}$ becomes $\overline{A\bar{B}}$

These four terms are directed into a NAND gate to produce:

$$\overline{\overline{\bar{A}B\bar{C}} \cdot \overline{\bar{B}C} \cdot \overline{AC} \cdot \overline{A\bar{B}}}$$

Check using DeMorgans

$$\overline{\overline{\bar{A}B\bar{C}} \cdot \overline{\bar{B}C} \cdot \overline{AC} \cdot \overline{A\bar{B}}} = \overline{\overline{\bar{A}B\bar{C}}} + \overline{\overline{\bar{B}C}} + \overline{\overline{AC}} + \overline{\overline{A\bar{B}}}$$

$$= \bar{A}B\bar{C} + \bar{B}C + AC + A\bar{B}$$

QED!

4 Number of gates

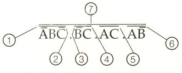

six two input NAND gates; one four input NAND gate; one 3 input NAND gate.

5 Circuit shown in Fig. 3.28.

Circuits for performing addition

Now is a good time to have a brief look at how combinational logic circuits can be used to perform common tasks. Digital systems operate using binary signals so if any mathematical operation is to be performed it uses binary

PRACTICAL INVESTIGATION 7

Implementing the Exclusive OR function

Show how the exclusive OR function can be implemented using (a) NAND gates only, and (b) NOR gates only.

Procedure: (a) NAND Gates only
1   Write the minimized Boolean expression $A \oplus B = \bar{A}B + A\bar{B}$.
2   Study the expression and specify the type of gates that are involved:

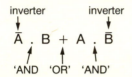

This gives two inverters, two AND gates and an OR gate.
3   Separate the expression into its main parts and analyse each part individually starting with the two AND gates.

$\bar{A}.B$, invert this term so that it forms a NAND gate $\overline{\bar{A}.B}$
$A.\bar{B}$, do the same with this term, $\overline{A.\bar{B}}$.
We now have: $\overline{\bar{A}.B} + \overline{A.\bar{B}}$. Now for the OR function that joins these terms.
    The rules say that to produce an OR function we must invert the inputs to a NAND gate. However, $\overline{\bar{A}.B}, \overline{A.\bar{B}}$ are already inverted so direct these straight into a NAND gate to give $\overline{\overline{\bar{A}.B}.\overline{A.\bar{B}}}$ this should be the answer.
    Check   using DeMorgans: $\overline{\overline{\bar{A}.B}.\overline{A.\bar{B}}} = \overline{\bar{A}.B} + \overline{A.\bar{B}} = \bar{A}.B + A.\bar{B}$.
4   Now count the NAND gates required

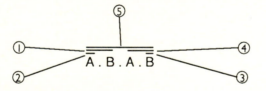

    five 2 input NAND are required QED!
5   Draw the resulting circuit Fig. 3.26.

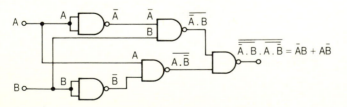

Fig. 3.26   Exclusive OR function (XOR) using NAND gates

Procedure: (b) NOR Gates only

Now let us implement the exclusive OR using NOR gates
1 $A \oplus B = \bar{A}.B + A.\bar{B}$.
2 Same gates as specified in the procedure for NAND gates.
3 $\bar{A}.B, A.\bar{B}$ are AND gates. The rules say invert the inputs to a NOR to make an AND function.
So $\bar{A}B$ becomes:

redundant term ⟶ $\overline{\bar{\bar{A}} + \bar{B}} = \overline{A + \bar{B}}$

likewise $A.\bar{B}$ becomes:

$\overline{\bar{A} + \bar{\bar{B}}} = \overline{\bar{A} + B}$

these terms are joined by an OR function which is a NOR followed by an inverter.
Now:

NOR ⟶ $\overline{\overline{(A + B)} + \overline{(A + \bar{B})}}$ this is it but check!

$\overline{\overline{(\bar{A} + B)} + \overline{(A + \bar{B})}} = \overline{(\bar{A} + B)} + \overline{(A + \bar{B})} = (\bar{\bar{A}}.\bar{B}) + (\bar{A}.\bar{\bar{B}}) = (A.\bar{B}) + (\bar{A}.B)$
4 Number of NOR gates required

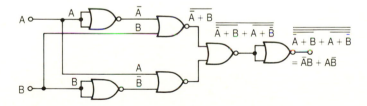

six two input NOR gates.
5 Resulting circuit shown in Fig. 3.27.

Fig. 3.27 Exclusive OR function (XOR) using NOR gates

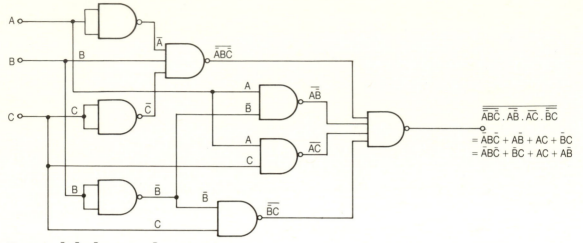

Fig. 3.28 $\bar{A}B\bar{C} + \bar{B}C + AC + A\bar{B}$ using NAND gates

arithmetic with binary addition following the rules:

$'1' + '0' = '1'$
$'0' + '1' = '1'$
$'1' + '1' = '0'$ carry $'1'$

We can now use our logic skills to design a Binary adding circuit.

A logic circuit adds together two input signals (A and B) to produce the sum; if a carry is involved this must also be indicated. The truth table is shown in Fig. 3.29. From the truth table

A	B	SUM	CARRY
0	0	0	0
0	1	1	0
1	0	1	0
1	1	0	1

Fig. 3.29   Simple adding circuit truth table

the Boolean expression can be derived.

$Sum = \bar{A}B + A\bar{B}$, carry $= AB$

These are both very familiar terms and can be performed by the circuit of Fig. 3.30. This circuit is known as a 'Half adder' because it is not able to handle any carry, that is the input from a previous stage or addition. *Note* The logic diagram and truth table show that the half adder can indicate when an overflow has occurred and an *output* 'carry 1' exists.

A 'Full adder' circuit will have the facility for adding any carry from a previous stage to the

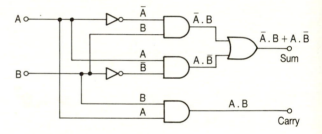

Fig. 3.30   Simple 'adder' circuit

sum of the inputs A and B. The truth table for this is shown in Fig. 3.31. This gives the following Boolean expression

$Sum = A\bar{B}\bar{C} + \bar{A}B\bar{C} + \bar{A}\bar{B}C + ABC$ (simplest form)

Carry out $= AB\bar{C} + A\bar{B}C + \bar{A}BC + ABC$ this minimizes to:

$AB + A\bar{B}C + \bar{A}BC + ABC$.

A	B	C(carry in)	SUM	CARRY OUT
0	0	0	0	0
1	0	0	1	0
0	1	0	1	0
0	0	1	1	0
1	1	0	0	1
1	0	1	0	1
0	1	1	0	1
1	1	1	1	1

Fig. 3.31   Truth table for a full adder

PRACTICAL INVESTIGATION *8*

Half adder using NAND gates

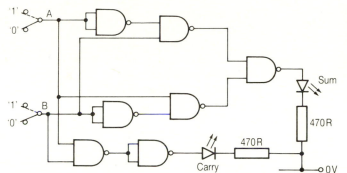

Equipment
Logic breadboard
2 × input switches
2 × LEDs
2 × 470R resistors
2 × quad 2 input NAND gates (4011 or 7400)
Power supply

Method
1 Build the circuit shown above after using the data sheets to determine the pin connections.
   NB   Remember to connect each chip to the power supply lines +5 V to pin 14, 0 V to pin 7.
2 By investigation complete the truth table shown below.

A	B	Sum	Carry
0	0		
0	1		
1	0		
1	1		

Results
1 Verify using DeMorgans rules that

i) $\overline{\overline{AB} \cdot \overline{A\overline{B}}} = \overline{A}B + A\overline{B}$

ii) $AB = \overline{\overline{AB}}$.

2 What would be the major disadvantage of making a 4 bit full adder (capable of adding two numbers each of 4 bits) from NAND gates only?

The logic circuit for a full adder is shown in Fig. 3.32.

Practical Investigation 8 allows you to implement a half adder circuit using NAND gates only and to test it fully.

If you have performed Practical Investigation 8 you will have found that a 2 bit half adder was quite complicated to wire up, so a 4 bit full adder using NAND gates only would be even worse! With lots of interconnections between

PRACTICAL INVESTIGATION **9**

The Full Adder

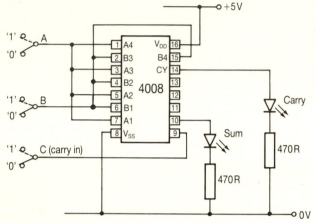

Equipment
Logic breadboard
3 × switches
2 × LEDs
2 × 470 R resistor
1 × 4 bit full adder (4008)
Power supply

Method

1   Build the circuit as shown above.
    Note   this circuit is a 4 bit full adder, which will enable a 4 bit number (A1 A2 A3 A4) to be added to another 4 bit number (B1 B2 B3 B4). We are using it as a simple adder to add input A to input B to produce a sum and a carry out. There is also an additional input C which represents a carry bit from a previous stage. For this circuit to work correctly all the A inputs (A1, A2, A3, A4) must be connected together and all the B inputs (B1, B2, B3, B4) must be connected together.

2   By investigation complete the truth table shown below

A	B	C carry in	Sum	Carry out
0	0	0		
1	0	0		
0	1	0		
1	1	0		
0	0	1		
1	0	1		
0	1	1		
1	1	1		

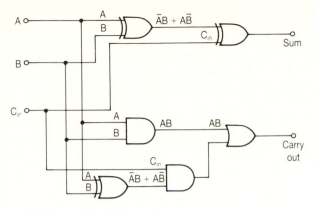

Fig. 3.32 Full adder circuit

gates and the possibility of many errors. The futility of building a full adder in this way becomes clear when you realize that complete circuits are already available as integrated circuits in both CMOS (4008) and TTL (7483) form. Practical Investigation 9 lets you test a purpose-built full adder. While you are wiring up this circuit take time to reflect upon just how long it would take if you were constructing it using individual NAND gates.

Small scale integration (SSI)

Integrated circuits of the type dealt with so far consist of a single package containing a number of gates. This is referred to as small scale integration (SSI), simply because the number of components contained within the silicon chip is small compared with what is achievable in other packages. As we have seen with the full adder, ICs are available that contain the equivalent of a number of interconnected gates, these are called medium scale integration (MSI) because they contain many more individual devices than their small scale counterparts.

When you have to implement a logic function or design a circuit it is worth looking to see if a purpose made IC is available that will perform the required task. If there is, this will undoubtedly provide the quickest, the most efficient and probably the most economical solution to the problem. The rule is: always check catalogues and data sheets before you embark on a design task.

Hazards

The output from a combinational logic circuit is determined by the state of the input variables. If an input should change from logic 0 to logic 1 and an unwanted output change occurs momentarily, a static hazard exists and a 'glitch' has occurred.

THOUGHT

Glitch! What's a Glitch? This is term used to describe any unwanted spike or pulse that occurs in a system. Consider the circuit of Fig. 3.33(a).

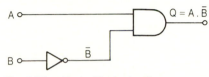

Fig. 3.33a Typical logic circuit

Chapter 2 showed that a propagation delay time (tpd) exists in all logic gates and this is the start of the problem. If you look at the timing of Fig. 3.33(b) you will see that when both inputs A and B are at logic 1 the output of the inverter should be logic 0. *But* because of the propagation delay of the inverter there will be a short period of time (tpd) when the output will be at logic 1 resulting in an unwanted output pulse or

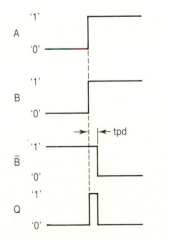

Fig. 3.33b Unwanted output pulse (glitch)

glitch. If a change in a single input can cause a glitch, a static hazard is said to exist. If there are a number of possible paths involved in the circuit, each introducing its own possibly different propagation delays, then a dynamic or race hazard exists.

Reducing the risk of hazards

When a logic function is to be minimized a Karnaugh map is used. The existence of a hazard shows up on the map if adjacent 1s are not interconnected in the same cell grouping. Study the map of Fig. 3.34(a). The grouping

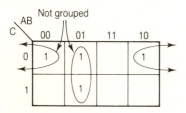

Fig. 3.34a   Original grouping

of this map gives $Q = \bar{B}\bar{C} + \bar{A}B$, but because there are adjacent 1s that are *not* grouped there is a hazard risk. If $\bar{C}$ and $\bar{A}$ are both 1 then:

$$Q = \bar{B}.\text{'1'} + \text{'1'}.B = \bar{B} + B$$

which we have already seen can cause a glitch.

Fig. 3.34(b) shows that by grouping these adjacent cells the hazard has been removed and the logic function is now:

$$Q = \bar{B}\bar{C} + \bar{A}\bar{C} + \bar{A}B.$$

Note   The term $\bar{A}\bar{C}$ is completely redundant, it is included in order to remove the hazard. Hazard free logic circuits are not often in minimized form.

When mapping hazard free circuits please remember that the top and bottom edges are

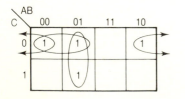

Fig. 3.34b   Hazard free grouping

considered adjacent, as are the sides, you may have to intergroup these edges to remove the risk of a glitch.

Combinational Logic Review

1   Any logic function can be produced by interconnecting the appropriate logic gates.
2   A truth table enables a complete Boolean expression for the output function to be derived.
3   Boolean output expressions can be derived from a logic circuit diagram.
4   The output expression for a logic function may be very long but its inverse may be short.
5   Long expressions may be minimized using:

   (a) Boolean algebra   (b) Karnaugh mapping.

6   DeMorgans laws are powerful tools for manipulating Boolean expressions:

   (i) $\overline{A+B} = \bar{A}.\bar{B}$ (ii) $\overline{A.B} = \bar{A}+\bar{B}$.

7   A Karnaugh map can be used to realize both the minimized function and the minimized inverse function.
8   When mapping, 'Don't care' or 'Can't happen' situations can be considered on the map as either '1' or '0' as appropriate in order to create the best grouping arrangement.
9   When designing practical logic circuits it is best to use gates of only one type, i.e. either NAND gates only or NOR gates only. This is called Universal logic.
10   Any gate function can be implemented using either NAND or NOR gates.
11   Complex circuits for performing mathematical functions, e.g. adders can be created using logic gates.
12   Standard logic gate integrated circuits contain relatively low numbers of individual devices and are known as small scale integration (SSI).
13   Integrated circuits containing complete circuits, e.g. Adders, multiplexers, decoders, etc., contain many individual devices and are called medium scale integration (MSI) ICs.

14 A complicated combinational logic circuit containing many gates may be available as a purpose built (MSI) integrated circuit.
15 Static and Race hazards may exist in a combinational logic circuit if an unwanted pulse (glitch) occurs due to the propagation delays of individual or collective gates.
16 Hazards can be minimized at the mapping stage by grouping adjacent cells to create additional groups.
17 Hazard free circuits generally involve redundant terms and hence require more gates.

Self Assessment Answers

Self Assessment 3

1

A	B	Q	
0	0	1	$\overline{A}\overline{B}$
0	1	0	
1	0	0	
1	1	1	AB

therefore $Q = \overline{A}\overline{B} + AB$.

2 If $\overline{A \oplus B} = \overline{A}\overline{B} + A\overline{B}$ then $Q = \overline{A}B + A\overline{B}$ must be the same as:

$Q = \overline{A}\overline{B} + AB$ (expression for exclusive NOR!)

So taking $Q = \overline{\overline{A}B + A\overline{B}}$

$Q = \overline{(\overline{A}B)}.\overline{(A\overline{B})}$

$= (\overline{\overline{A}} + \overline{B}).(\overline{A} + \overline{\overline{B}})$

$= (A + \overline{B}).(\overline{A} + B)$

$= A\overline{A} + AB + \overline{B}\overline{A} + \overline{B}B$

BUT $A\overline{A} = $ '0' and $B\overline{B} = $ '0'

so $Q = AB + \overline{A}\overline{B}$.

Self Assessment 4

1 Use Boolean algebra to minimize the expression

$Q = ABC + AB\overline{C} + A\overline{B}C + \overline{A}BC$

using the logic identity $A + A = A$
∴ $ABC + ABC = ABC$

$Q = ABC + AB\overline{C} + ABC + A\overline{B}C + ABC + \overline{A}BC$

the term ABC is introduced twice to make grouping of terms and factorizing easier.

$Q = AB(C + \overline{C}) + AC(B + \overline{B}) + BC(A + \overline{A})$

the logic identity $A + \overline{A} = $ '1'

$Q = AB + AC + BC$.

Self Assessment 5

1

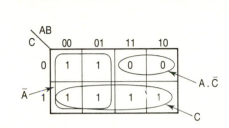

(a) $Q = \overline{A} + C$
(b) Grouping the empty cells gives $\overline{Q} = A.\overline{C}$.
2 Draw the map and plot the inverse Q terms as logic '0'.

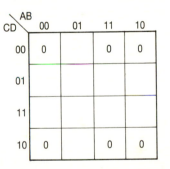

Now write logic '1' in all the empty cells and then group and minimize to give the Q function.

$Q = \overline{A}B + D$.

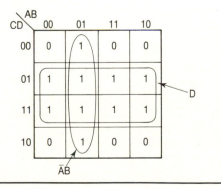

4
Sequential Logic

'All sequential logic circuits rely on the existence of a basic memory or storage device, usually a bistable.'

The bistable

Often called the 'flip-flop' this is a device that has two stable states. If a flip-flop is triggered so that its output is say logic 1 it will remain in this state until it is made to change, i.e. it has a memory.

A glance at a catalogue or set of data sheets will reveal the type of flip-flops available, there is a bewildering array.

Flip-flops

These fall into a number of categories: the RS type; the D type; the T type; the JK type; the master-slave JK type; and the Edge triggered type. In order to help gain an understanding of the various types of flip-flop it is a good idea to take a very brief look at the operation of the basic type, the RS (or SR) bistable.

The RS flip-flop

The basic logic circuit using NOR gates is shown in Fig. 4.1(a) with a block diagram in Fig. 4.1(b). The truth table is shown in Fig. 4.2. Note that the state of the output is determined by the inputs, but when both inputs are at Logic 0 the previous state is held — or stored. The situation where both inputs can be at Logic 1 results in an indeterminable output (impossible to predict).

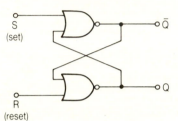

Fig. 4.1a   NOR gates RS flip-flop

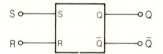

Fig. 4.1b   Symbol for NOR gate RS flip-flop

S	R	Q	$\bar{Q}$	
1	0	1	0	'1' on S sets Q to '1'
0	1	0	1	'1' on R sets Q to '0'
0	0	←		Outputs remain in their previous state
1	1	←		NOT ALLOWED (indeterminable)

Fig. 4.2   Truth table for NOR gate RS flip-flop

An RS flip-flop can also be made using NAND gates as shown in Fig. 4.3(a) and 4.3(b).

Note   The circuit symbol has circles drawn at the inputs to show this device triggers from a Logic 0, i.e. it is active low and is thus the opposite to the NOR RS type.

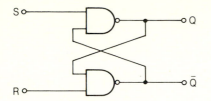

Fig. 4.3a NAND gate RS flip-flop

Circles indicate 'negated'
(or active low) inputs

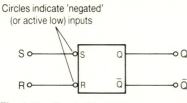

Fig. 4.3b Symbol for NAND gate RS flip-flop

THOUGHT

*Does this mean that to set Q to '1', S must be '0'? Yes.
Study the table of Fig. 4.4.*

S	R	Q	$\bar{Q}$	
0	1	1	0	'0' on S sets Q to '1'
1	0	0	1	'0' on R sets Q to '0'
1	1	←		Last state held
0	0	←		NOT ALLOWED (indeterminable)

Fig. 4.4 NAND gate RS flip-flop truth table

These devices are both available in CMOS form: 4043 = NOR type, 4044 = NAND types (but are called RS latches which is an alternative name).

The clocked RS flip-flop

The previous chapter mentioned the problems that can be caused by a static or dynamic hazard. To recap briefly, this is where the individual propagation delays of interconnected gates can cause a problem. As input data is changing there may be a brief instant when an unsolicited output pulse or glitch is produced. This could result in incorrect or unreliable flip-flop operation. A solution to this problem is to ensure that the Set and Reset operations can

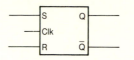

Fig. 4.5 Clocked RS flip-flop

only occur at specified times. This is achieved by providing the flip-flop with a clock input, as shown in Fig. 4.5. The flip-flop will now only change state when a clock pulse is received; if a pulse is not received the flip-flop output will not change. The waveforms of Fig. 4.6 show the operation of a clocked RS flip-flop.

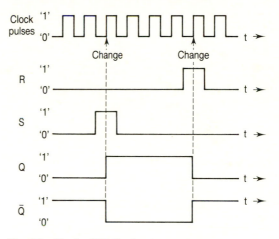

Fig. 4.6 Clocked RS flip-flop (no propagation delay shown)

Notice that the output Q does not change to '1' as soon as S becomes '1' but only when the clock pulse changes to '1'. Likewise if the reset changes to '1' the Q output changes to '0' when the next clock pulse is received.

The D type flip-flop

Clearly the indeterminable state (S = R = '1') of the RS flip-flop can lead to problems. A solution here is to provide the clocked RS flip-flop with a single input as shown in Fig. 4.7(a) and 4.7(b). You will see that the output Q will take on the state of the D input upon receipt of the next clock pulse as shown in Fig. 4.8.

Important facts about clock pulses

From what has been said so far it is apparent that the clock pulse is an all important operating

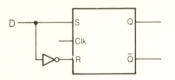

Fig. 4.7a   D type flip-flop

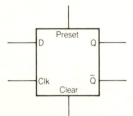

Fig. 4.7b   Symbol for D type flip-flop

D	Clock	Q	$\bar{Q}$
1	⎍	1	0
0	⎍	0	1

Fig. 4.8   Truth table for clocked D type flip-flop

signal. From the discussions of Chapter 1 a pulse has the following characteristics:

1   it has a leading or rising edge;
2   it has a trailing or falling edge;
3   a level change occurs, i.e. from '0' to '1' back to '0' or from '1' to '0' back to '1'; this means that a flip-flop could be triggered by:
   (a)   a level change;
   (b)   the rising edge;
   (c)   the falling edge.

In practical terms this means that if the data is entered into the flip-flop by a specified change in clock pulse level, it is a level triggered device. If however the data is entered at the time the clock pulse changes level, it is being triggered by the edge of the pulse and is thus an edge triggered flip-flop.

THOUGHT

So a flip-flop can be triggered by a positive (active high) or negative (active low) clock pulse level, or be triggered by the positive going or negative going edge of a pulse. How do you know which it is? Happily a convention exists that uses the symbols shown in Fig. 4.9. The wedge symbol indicates edge triggering and the tiny circle indicates negation or inversion, i.e. triggered by a logic 0 *not* a logic 1.

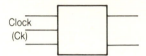

Fig. 4.9a   Positive (high) level triggered

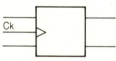

Fig. 4.9b   Negative (low) level triggered

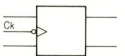
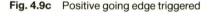

Fig. 4.9c   Positive going edge triggered

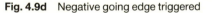

Fig. 4.9d   Negative going edge triggered

It is probably fair to say that the majority of modern logic circuits employ the use of edge triggered flip-flops. This is because the edge of a pulse has a very short time duration and so the possibility of false operation due to noise and interference is reduced.

The JK flip-flop

This is a similar device to the RS flip-flop except that it has additional circuitry that takes care of the indeterminable states. The logic symbol is shown in Fig. 4.10(a) with its associated truth table Fig. 4.10(b).

This truth table is normally shown in the format of Fig. 4.10(c). This shows the 'shorthand' terms: Qn to denote Q, $Qn+1$ to denote Q after one clock pulse and $\overline{Qn}$ to denote a change in Q.

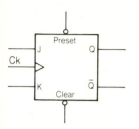

Fig. 4.10a   Logic symbol for JK flip-flop

J	K	Q	$\overline{Q}$	
0	0	No change		Q = What it was!
0	1	0	1	Q reset to '0' after clock pulse
1	0	1	0	Q set to '1' after clock pulse
1	1	Toggle		Q changes from its previous state with every clock pulse

Fig. 4.10b   JK truth table

J	K	Qn+1	(Q after 1 clock pulse)
0	0	Qn	Q remains in its last state (denoted by Qn)
0	1	0	(Q reset to '0')
1	0	1	(Q set to '1')
1	1	$\overline{Qn}$	(Q changes from its previous state with every clock pulse)

Fig. 4.10c   Conventional JK truth table

Master-slave flip-flops

The disadvantage with level triggered flip-flops is that they are prone to Race hazards. If the J and K input data should change during the time that a clock pulse is present the output will change also. To avoid this unwanted reaction

the input data must be carefully conditioned so that it cannot change during the duration of the clock pulse. The practical problems created by this situation are difficult to solve in a satisfactory manner, and this has led to the development of the master-slave flip-flop shown diagrammatically in Fig. 4.11.

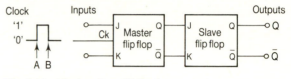

Fig. 4.11   Master-slave JK flip-flop

This flip-flop consists of two flip-flops! When the clock pulse is at Logic 0 the inputs are connected to the input of the master flip-flop. When the clock pulse changes from '0' to '1' (point A on the waveform) the inputs are isolated from the master and its Q outputs are passed on to J and K inputs of the slave flip-flop. When the clock pulse changes from '1' to '0' (point B on the waveform) the slave flip-flop outputs change. The net result of this is that a master-slave flip-flop's output changes with the negative going edge of the clock pulse. This ensures that the data proceeds in an orderly fashion through the device and the possibilities of a race hazard occurring are avoided.

The Edge triggered flip-flop

This behaves just like the master-slave device but the circuitry inside the IC is different. The output will change state with the edge of the clock pulse. With a negative edge triggered device it is the falling edge of the clock pulse Fig. 4.12(a) ('1' to '0' transition) that shifts the data to the output terminals — exactly like the

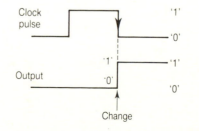

Fig. 4.12a   Negative going edge triggering

PRACTICAL INVESTIGATION 10

Flip-flop Triggering

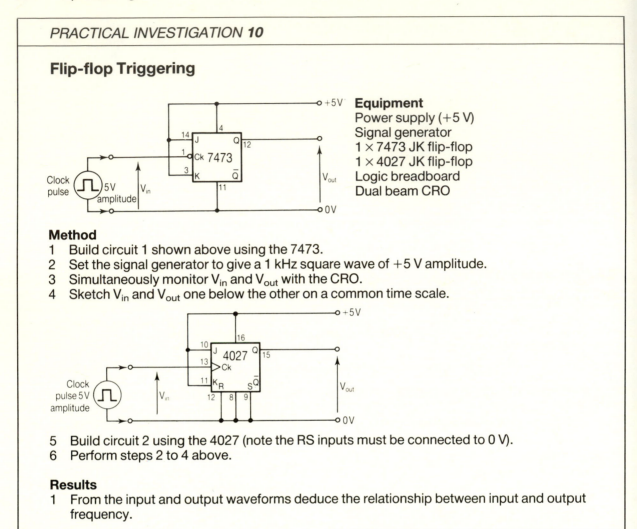

Equipment
Power supply (+5 V)
Signal generator
1 × 7473 JK flip-flop
1 × 4027 JK flip-flop
Logic breadboard
Dual beam CRO

Method
1 Build circuit 1 shown above using the 7473.
2 Set the signal generator to give a 1 kHz square wave of +5 V amplitude.
3 Simultaneously monitor V_{in} and V_{out} with the CRO.
4 Sketch V_{in} and V_{out} one below the other on a common time scale.

5 Build circuit 2 using the 4027 (note the RS inputs must be connected to 0 V).
6 Perform steps 2 to 4 above.

Results
1 From the input and output waveforms deduce the relationship between input and output frequency.

master slave type. A positive going edge triggered flip-flop is operated by the rising edge of the clock pulse as indicated in Fig. 4.12(b).

Practical Investigation 10 allows you to compare a negative going edge triggered TTL JK flip-flop (7473) with a positive going edge triggered CMOS type (4027).

THOUGHT

Should sequential logic circuits use gates with different triggering characteristics? No! In order to prevent race hazards one type of triggering should be employed. The majority of modern circuits employ edge triggered devices rather than level triggered types.

The T type flip-flop

If you performed Practical Investigation 10 you will have discovered that for a JK flip-flop when

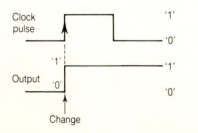

Fig. 4.12b   Positive going edge triggering

J = K = '1' the output changes state with every clock pulse — i.e. it 'toggles'. This means the device is operating as a toggle or T type flip-flop. You will see that this is a 'divide by 2' device since the output frequency will be half that of the input clock frequency.

Practical Sequential Circuits

From the facts that have been presented up to this point we can make a number of statements about a flip-flop:

1 The inputs will determine the state of the output (Q).
2 There is a complementary or inverse output (Q̄).
3 The state of the output will only change when a clock pulse is applied to the circuit.
4 There are facilities for setting (S) and resetting (R) the flip-flop.
5 Flip-flops are obtainable with either active high ('1') or active low ('0') inputs.
6 A flip-flop is a bistable and hence is a divide by 2 circuit.

Flip-flops can be interconnected to create counters or shift registers that are activated by clock pulses. We shall now explore these circuits.

A further word about clock pulses

It is vital that the clock pulses which are used to operate sequential logic circuits are clean with fast rise and fall times. However in order to observe the operation of sequential circuits they must be clocked slowly, i.e. one pulse at a time. A switch is eminently suitable for this but it must be bounce free or debounced. Please ensure that any clock pulses for the following Practical Investigation are supplied via a debouncing circuit, the one shown in Fig. 4.13 is ideal since it uses a LED to indicate when the output pulse is high (Logic 1).

You may wish to pulse some circuits quickly so it might be a good idea to build a low

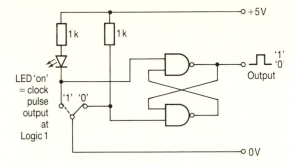

Fig. 4.13   Debounced switch using NAND gates

frequency pulse generator using the 555 Timer circuit of Fig. 4.14.

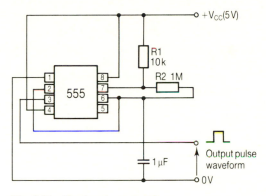

Fig. 4.14   Clock pulse generator

Before you commence work building and testing sequential logic circuits you should have the following:

4 switches A, B, C, D for providing input data,
1 debounced switch for providing the clock pulses
a low frequency (0.5 Hz) logic pulse generator for providing a train of clock pulses (this is optional).

If you build these on the breadboard where you intend to construct your circuit there will not be much room left. As suggested in the introduction therefore, build the switching and clocking circuits on one breadboard and then connect such outputs as you need to a second breadboard and use this as your construction site. One version of this set up is shown in Fig. 4.15.

PRACTICAL INVESTIGATION 11

4 stage Asynchronous Counter

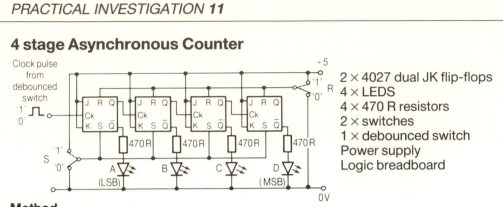

2 × 4027 dual JK flip-flops
4 × LEDS
4 × 470 R resistors
2 × switches
1 × debounced switch
Power supply
Logic breadboard

Method

1   Build the circuit shown above.
2   Set the R switch to '0'.
3   Set all stages by moving the switch S to '1' and back to '0'.
4   Apply clock pulses with the debounced switch and observe the output LEDs.
5   Continue 'clocking' this circuit while you observe its operation.
6   Apply a number of clock pulses and then set the circuit with switch S (move to '1' and back to '0').
7   Commence clocking again and reset the circuit with R (move to '1' and back to '0').
8   Connect the LEDs to the Q output of each stage and repeat steps 2 to 7.
9   If you have made a pulse generator, connect this in place of the debounced switch input and observe the operation.

Results

1   What is the purpose of the Set and Reset lines?
2   What are the disadvantages of this type of counter?

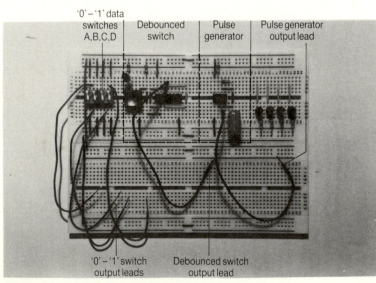

Fig. 4.15   Logic breadboard layout

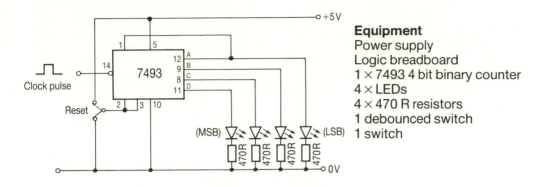

PRACTICAL INVESTIGATION 12

Four Stage Binary Counter (1)

Equipment
Power supply
Logic breadboard
1 × 7493 4 bit binary counter
4 × LEDs
4 × 470 R resistors
1 debounced switch
1 switch

Method
1 Study the pin connection diagram for the 7493 IC and then connect it to make the circuit shown in schematic form above.
2 Supply clock pulses from a debounced switch and note how the outputs A to D change. *Note* that the device is triggered by high to low ('1' to '0') transitions of the clock pulse.
3 If you have built a low frequency clock pulse generator use this in place of the single clock pulse switch and observe the operation.
4 Do not dismantle this set-up but save it for the next investigation.

Asynchronous Binary Counters

This is often called a ripple through counter, and can be constructed by connecting flip-flops in series as illustrated by the four stage counter of Fig. 4.16. Notice that the J and K input of every flip-flop is permanently held at '1' so they are all in toggle mode and will each divide by 2.

Flip-flop A changes state after every clock pulse, B after every two, C after every four and D after every eight. The result of this is that the four outputs (A to D) represent the number of clock pulses in binary form. This produces a

counter that will count from zero to fifteen and then reset to zero on the sixteenth pulse. The operation of this counter is explained best by Practical Investigation 11.

The asynchronous counter clearly works well enough but it suffers from a major disadvantage. Each flip-flop is triggered by the previous stage, when the propagation delay of each stage is considered it may be seen that the maximum operating frequency will be limited.

For example suppose in a four flip-flop system each flip-flop introduced a propagation delay of 50ns. The overall propagation delay will be 4 × 50ns = 200ns. This means that the

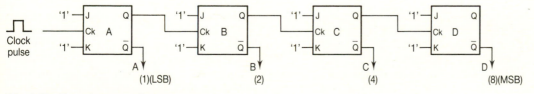

Fig. 4.16 Four stage asynchronous counter

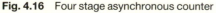

PRACTICAL INVESTIGATION *13*

Four Stage Binary Counter (2)

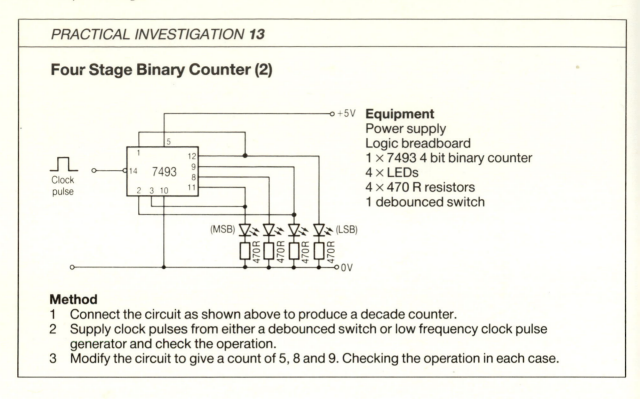

Equipment
Power supply
Logic breadboard
1 × 7493 4 bit binary counter
4 × LEDs
4 × 470 R resistors
1 debounced switch

Method

1   Connect the circuit as shown above to produce a decade counter.
2   Supply clock pulses from either a debounced switch or low frequency clock pulse generator and check the operation.
3   Modify the circuit to give a count of 5, 8 and 9. Checking the operation in each case.

maximum clock pulse rate (prf) or operating frequency will be limited to:

$$\frac{1}{200 \times 10^{-9}} = 5 \text{ MHz.}$$

Provided that this limitation is observed this type of counter will operate reliably. However the circuit uses four JK flip-flops and is quite tedious to build, imagine what it would be like if you wanted a 6, 7, or 14 bit counter! The last chapter mentioned that medium scale integration (MSI) allows complete circuits to be fabricated in a single package, so rather than build these from individual devices it is sensible to buy them ready made. Asynchronous counters include the TTL 7493 (a 4 bit counter), the CMOS 4024 (a 7 bit counter) and the 4020 (a 14 bit counter).

Perform Practical Investigation 12 and see how a purpose made MSI 4 bit counter compares with the four flip-flop version.

If you performed Practical Investigation 11 you will appreciate how much easier it is to use the ready made 4 bit counter as opposed to building it from four flip-flops.

The device contains four master-slave type flip-flops with each output supplying the clock pulse to the next one in the line in ripple through fashion. Notice from the pin out diagram of the 7493 (page 115) that the output of the first stage (A) is not connected to the other stages, but provides a divide by 2 facility. To operate as a 4 bit counter, output A must be connected to the clock input (pin 1) of the next stage.

Reducing the count!

The 4 bit counter will count pulses from 0 to 15 giving a count of 16, (2^4). This is fine if you want to count to 15 and then reset to zero but supposing you wish to count to 10 or 8 or 12 and then reset to zero. You will notice by studying the circuit of Practical Investigation 12 that the reset facility is via a two input AND gate. When both inputs to this gate are Logic 1 the circuit resets to zero. To make a counter reset after a given number of input pulses it is only necessary to select the appropriate flip-flop outputs and connect them to the reset AND gate. Suppose a decade counter is required, this must

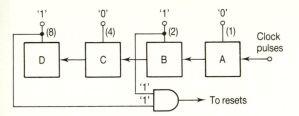

Fig. 4.17   Decade counter

count from 0 to 9 and reset on the count of 10. The diagram of Fig. 4.17 shows the state of the flip-flop outputs on the count of 10. You will see that output D is high since it represents denary 8, and so is output B since it is denary 2. If these outputs are connected to the reset gate it will set the counter to zero on the count of 10. This principle can be applied to give any desired count within the counter's operating range as Practical Investigation 13 will enable you to ascertain.

Synchronous Counters

While the asynchronous counter operates in a satisfactory manner it does take time for the pulse to ripple through as it toggles each flip-flop, also if it is operated at or near its maximum frequency there is a risk that a hazard or glitch will occur. The speed of operation can be greatly increased if all the flip-flops in the counter are clocked together, i.e. synchronously operated. Consider the circuit of the three stage counter shown in Fig. 4.18.

Notice that each flip-flop is a T type and that the output (Q) of each stage provides the input to the next stage. The clock pulses are applied simultaneously to all stages. Each input will be at Logic 1 when both the input and output of the previous stage is at Logic 1. The timing diagram for this circuit is shown in Fig. 4.19.

All stages are clocked together but there will be a small difference between the toggle time of each stage. This is further increased by the propagation time of the associated AND gate. The maximum clock rate will thus be limited by the propagation delay of a pair of the JK flip-flops and the AND gate.

It would be possible to build a synchronous counter using separate flip-flops and associated gates, but it is cheaper and quicker to use the readily available MSI devices that exist. It is possible to obtain synchronous counters that provide a variety of functions. Have a look at the selection shown below and then carry out the related Practical Investigations 14 and 15.

TTL Synchronous Counters
74161 4 bit counter with direct clear
74162 4 bit BCD decade counter
74163 4 bit binary counter with synchronous clear
74169 4 bit up/down counter
74193 4 bit up/down with dual clock
74160 binary coded decimal (BCD) counters
74168 BCD up/down counter
74192 BCD up/down with dual clock.

CMOS Synchronous Counters
4029 up/down binary/BCD counter
40161 4 bit counter
40163 4 bit counter
40193 4 bit up/down counter with separate clocks
40160 BCD counter
40162 BCD counter
40192 BCD up/down separate clocks.

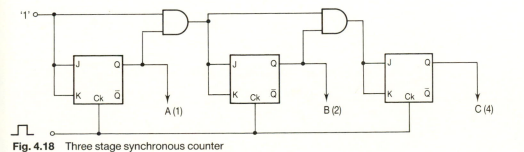

Fig. 4.18   Three stage synchronous counter

PRACTICAL INVESTIGATION *14*

Synchronous Binary Counter

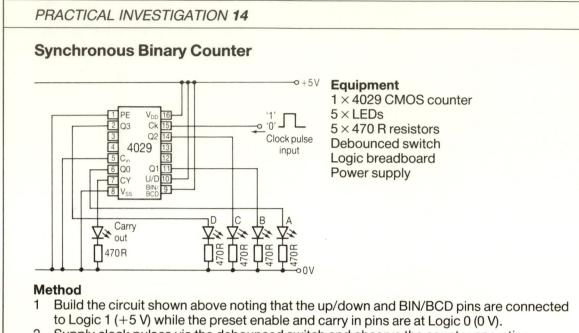

Equipment
1 × 4029 CMOS counter
5 × LEDs
5 × 470 R resistors
Debounced switch
Logic breadboard
Power supply

Method
1   Build the circuit shown above noting that the up/down and BIN/BCD pins are connected to Logic 1 (+5 V) while the preset enable and carry in pins are at Logic 0 (0 V).
2   Supply clock pulses via the debounced switch and observe the counter operation, paying particular attention to the logic state of the carry out terminal.
3   Connect the up/down pin to Logic 0 (0 V) and repeat the above procedure.

Results
1   Does the counter increment on the positive or negative edge of the clock pulse?
2   When counting up, what is the logic state of the carry out:
    (a)   During the count?   (b)   When the count is maximum?
3   When counting down, what is the logic state of the carry out:
    (a)   During the count?   (b)   When the count is minimum?

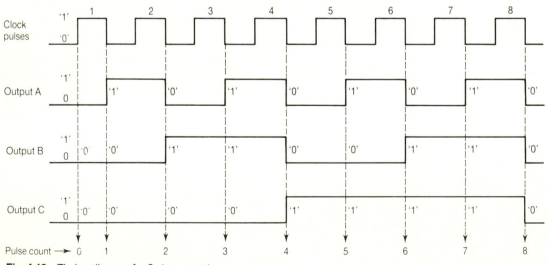

Fig. 4.19   Timing diagram for 3 stage synchronous counter

PRACTICAL INVESTIGATION *15*

Synchronous BCD Counter

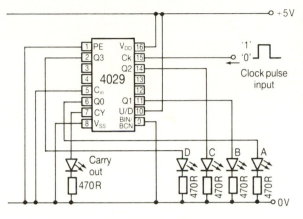

Equipment
1 × 4029 CMOS counter
5 × LEDs
5 × 470 R resistors
Debounced switch
Logic breadboard
Power supply

Method
1 Build the circuit shown above noting that the up/down pin is connected to Logic 1 (+5 V), while the preset enable, carry in and BIN/BCD pins are at Logic 0 (0 V).
2 Supply clock pulses via the debounced switch and observe the counter operation.
3 Connect the up/down pin to Logic 0 (0 V) and repeat the above procedure.
4 If you have a low frequency pulse generator use this to supply the clock pulses and observe the counter operation.

Shift Registers

A shift register is basically a number of flip-flops connected in series or cascade as shown in Fig. 4.20. Note that the first JK flip-flop is connected as a D type. In this circuit data is shifted through the register by the negative going edge of the clock pulse. Note that the flip-flops are clocked together (synchronously) and that all the resets are interconnected, making it possible to clear all the devices at the same time. Notice

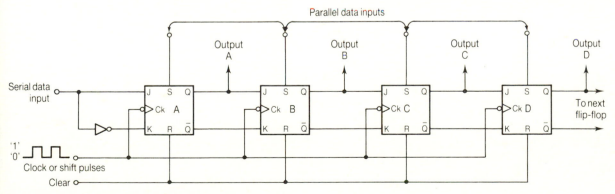

Fig. 4.20 A 4 bit shift register using JK flip-flops

also that it is possible to apply data into the first flip-flop so that it can be passed along the line of flip-flops to the output. Alternatively data can be applied to each individual flip-flop at the same time via the set or preset inputs. This is parallel operation. Let us now look at how the circuit works from both aspects.

Consider the case where all the units are cleared and the input data is applied to the J input of flip-flop A in the form of a pulse train as shown in Fig. 4.21(a). If you study the output table of Fig. 4.21(b) you can see that the clock

MSB A	B	C	LSB D	Clock pulse number
0	0	0	0	No pulses
1	0	0	0	⌐Ł 1st
0	1	0	0	⌐Ł⌐Ł 2nd
1	0	1	0	3rd = data shifting right
1	1	0	1	4th = data loaded
0	1	1	0	5th = data right
0	0	1	1	6th = data right
0	0	0	1	7th = data right
0	0	0	0	8th = data completely out, register empty

Shift →

(b)

'1' '1' '0' '1'

MSB   LSB

Serial input data = 1101

(a)

Fig. 4.21   Shifting action of 4 bit register

pulse shifts the input data right one bit at a time through the register. It takes four clock pulses to load the register with the input data and a further four pulses to shift this data to the output one bit at a time. There are some important points to be noted here:

1   The input data is in serial form, i.e. a train of pulses one behind the other making up the data word.
2   The negative edge of each clock pulse shifts this data one place to the right into the register.
3   After four clock pulses the register holds the input data word. The outputs A, B, C, D

indicate this word and since they are available together the device has a serial input and a parallel output.
4   Further clock pulses continue to shift the data to the right where it appears at the output one bit at a time. This means that the circuit also provides a serial input and a serial output function.

THOUGHT

With a serial in/serial out function the input data does not appear at the output until eight pulses have occurred. Does this mean it is delayed? Yes it does, one application of this type of shift register is to use it as a delay circuit.

Let us now consider how the set or preset inputs can be used. Remember a Logic 1 on the 'S' input to a JK flip-flop will set the Q output to Logic 1. A data word can be loaded into the register by the appropriate logic signals on the S input, or parallel data inputs as they are called, in Fig. 4.20.

Consider that the data word: 1 1 0 1 is loaded into the register using the parallel data inputs.

To do this the S input must be at Logic 1 for flip-flops A, B and D while the S input of flip-flop C must be Logic 0. The outputs A, B, C, D will now indicate this data word. If clock pulses are now applied it will take four to shift this data word to the output where it appears one bit at a time. The circuit is now operating as a parallel input, serial output shift right register.

As with other circuits, although shift registers can be built using flip-flops they are readily available as a single MSI device. These offer the facilities of shifting serial or parallel data right or left and providing parallel or serial outputs according to the type. Study the small selection shown below and then prove for yourself the operation of a shift register with Practical Investigations 16 and 17.

TTL Shift Registers

Code no.	Facility	Clock pulse triggering
7495	4 bit serial/parallel input parallel output; shift left/ shift right.	negative going edge
7496	5 bit serial/parallel right, parallel output.	negative going edge

PRACTICAL INVESTIGATION *16*

Parallel input/Parallel output Shift Register

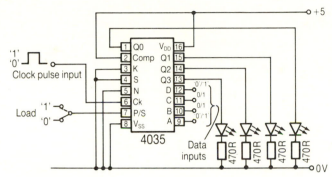

Equipment
1 × CMOS 4035 shift register
Power supply
Logic breadboard
Debounced switch
4 × LEDs
4 × 470 R resistors

Method
1   Connect the circuit as shown above.
2   Notice that there are four parallel data inputs (A, B, C, D) and four outputs (Q0, Q1, Q2, Q3). Enter the data word 1100 into the register by connecting the appropriate input to Logic 1 (+5 V) and Logic 0 (0 V). E.g. A = +5 V, B = +5 V, C = 0 V, D = 0 V.
3   With the load switch set to Logic 0 operate the clock pulse switch several times — observe that nothing happens.
4   Set the load switch to Logic 1 and operate the clock pulse switch several times.
5   Now set the load switch to Logic 0 and operate the clock pulse switch.
     Hopefully you will have found that the data is loaded into the register by a Logic 1 on the load input and one clock pulse. Further clock pulses do not change the output. When the load is set to Logic 0, data in the register is shifted by each clock pulse.
6   Change the input data word by making the appropriate connections and investigate the shifting operation. If you have a low frequency clock pulse generator use this to clock the register.
7   During a clocking sequence note the effect of setting the R input to Logic 1 and back to Logic 0.
8   Repeat the investigation with the 'comp' (pin 2) input connected to Logic 0.

Results
1   What is the purpose of the R input?
2   What is the purpose of the comp input?
3   Is this shift register operated by positive or negative going transitions of the clock pulse?

74164	8 bit, serial input parallel output.	positive going edge	
74165	8 bit serial/parallel input, serial output.	positive going edge	
74166	8 bit serial /parallel input, serial output, universal shift register.	positive going edge	
74194	4 bit, bidirectional serial/ parallel input/parallel output.	positive going edge	
74198	8 bit, universal shift register.	positive going edge	

PRACTICAL INVESTIGATION *17*

A Universal Shift Register

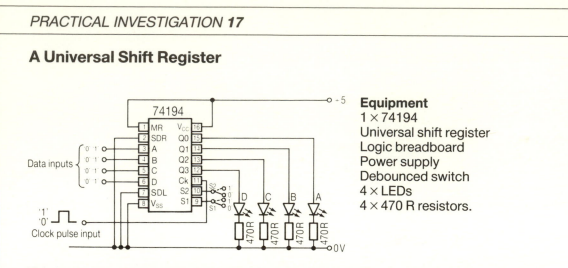

Equipment
1×74194
Universal shift register
Logic breadboard
Power supply
Debounced switch
$4 \times$ LEDs
4×470 R resistors.

This is a universal shift register and as such will shift parallel or serial input data left or right according to how it is operated.
Note   The reset line (MR) resets all the outputs (Q0, Q1, Q2, Q3) to Logic 0 when MR is set to Logic 0. There are four parallel data inputs (A, B, C, D) and two serial data inputs:
SDL = Serial data input left shift   SDR = Serial data input right shift.
The data are selected (or loaded according to the state of the S1 and S2 inputs).

Mode	S1	S2
Parallel load	'1'	'1'
Hold	0	0
Shift left	1	0
Shift right	0	1

Let us now proceed with the investigation.

Parallel input/parallel output

Method
1   Build the circuit shown above.
1   Connect the parallel data inputs to give the data word
    1 1 0 0 (A = +5 V, B = +5 V, C = 0 V, D = 0 V).
    2   Set S1 and S2 to Logic 1 and input one clock pulse.
4   Operate the clock pulse switch several times and observe that the output data remains unchanged (at 1 1 0 0).
5   Set S2 to Logic 0 and provide two clock pulses observing the data shift.
6   Set S2 to Logic 1 and S1 to Logic 0 and provide two clock pulses observing the data shifts back.
7   Investigate the shifting of the parallel data by:
    (a)   shifting left and right;
    (b)   changing the input data word (remember that to load the new data both S1 and S2 must be at Logic 0);
    (c)   noting the effect of changing S1 and S2 to Logic 1;
    (d)   noting the effect of setting the MR input to Logic 0.

Serial input/parallel output

Method

1 Set the parallel input data to 0 0 0 0
   (A = B = C = D = Logic 0).
2 Set S1 to Logic 1, S2 to Logic 0.
3 Set SDL to Logic 0.
4 Use SDR as the serial data input and perform the following:
   (a) Set SDR to Logic 1 and input two clock pulse.
   (b) Set SDR to Logic 0 and provide a further four pulses observing the data shift through
       the register.
5 Set S1 to Logic 0 Set to Logic 1. Set SDR to Logic 0.
6 Use SDL as the serial input and repeat steps 4(a) and (b).
7 Investigate fully the operation of the register by experimenting with the various input
   permutations for both serial and parallel input operation.

CMOS Shift Registers

4014	8 bit parallel input, serial output.	positive going edge
4015	Dual 4 bit, serial input, parallel output.	positive going edge
4021	8 bit, parallel input, serial output.	positive going edge
4035	4 bit parallel input, parallel output.	positive going edge
40194	4 bit bidirectional parallel input, parallel output.	positive going edge

Note that this is a small selection but many others are available.

Logic circuits other than counters and shift registers

The flip-flop forms the basis for the counter and shift register, and we have seen how these complete circuits can be obtained as a medium scale integrated circuit. There are a number of inexpensive commercial MSI circuits that provide specialized functions in a single dual in line (d.i.l.) package. These include code converters and multiplexers and demultiplexers.

Encoders and Decoders

Since logic circuits operate by using only two levels, the input and output signals are binary. However there is often a need to convert one type of binary code into another, or to convert a binary code into a decimal code and vice versa. In addition to these commonly used codes there are a number of other codes like XS3 (excess three), octal (base 8), hexadecimal (base 16) and Gray that may require converting. It is possible to provide code conversion using logic circuits and in particular purpose made MSI chips. To consider this aspect let us look at the simple codes that are most frequently used.

Binary Code

This is the basic code consisting of a numbering system to the base 2. Each binary digit or bit represents a power of 2 so the binary number 1 0 0 1 1 0 1 1 can be shown to have a decimal or denary value of 155. This is also known as positional binary since the position which each bit occupies in the binary number reflects its particular value so.

Thus $\qquad$ 1 0 0 1 1 0 1 1

becomes $\qquad 128 + 16 + 8 + 2 + 1 = 155_{10}$

Using this numbering system arithmetic operations can be carried out in a similar way to those for denary numbers.

Binary Coded Decimal (BCD)

A four bit binary number can be used to represent a denary number from 0 to 15 since $0000 = 0_{10}$ and $1111 = 15_{10}$. However since you performed Practical Investigation 15 you will know that a four bit binary word can be used to represent a denary number from 0 to 9 if it resets to 0 after 9.

By using four bits to represent each digit of a decimal number large numbers can easily be interpreted in binary form. Thus the number decimal 2597 can be represented as:

0010 0101 1001 0111

  ↓   ↓   ↓   ↓

  2   5   9   7

This requires the use of 16 binary digits. A BCD number will always contain more bits than the positional binary code representation. Consider the number 26:

Using positional binary code $26_{10} = 11010$

(5 bits)

Using BCD $26_{10} = 0010\ 0110$

(8 bits)

     ↓   ↓

     2   6

There is a need for BCD because we are often required to enter denary numbers into a binary logic circuit and then display the binary output as a denary number. More about this in the next chapter.

Octal Code

This is a number system to a base of 8; this means that the counting is from 0 to 7. Octal code can be used to represent binary numbers so that they are easier to handle. For example, consider the binary number 010011011. This represents the denary number 155 but is hardly instantly recognizable. If octal code is used then 3 binary 'bits' are required to represent numbers up to and including 7 as shown below.

Octal	Binary
0	000
1	001
2	010
3	011
4	100
5	101
6	110
7	111

By taking binary numbers in groups of three,

their octal equivalent can be found:

010 011 011

 ↓   ↓   ↓

 2   3   3

The long binary number 010011011 can thus be represented by 233_8, which is its octal equivalent.

Note   The subscript 8 is used to indicate that it is an octal number.

THOUGHT

So the denary number 155 is represented in binary as 010011011 and in octal as 233_8? Quite so, thus $155_{10} = 233_8$. In this way long and unwieldy binary numbers can be represented in a compact form that it is easy to convert, e.g.

$010101110111 = 2567_8$

$110011 = 63_8$

$1000011 = 103_8$

Note   that the grouping commences from the LSB. Likewise:

$$503_8 = 5 \qquad 0 \qquad 3$$
$$= 101 \qquad 000 \quad 011$$
$$= 5 \times 8^2 + \ 0 \times 8^1 + 3 \times 8^0$$
$$= 5 \times 64 + 0 \times 8 \ + 3 \times 1$$
$$= 320 \ + \ 0 \ \ + \ 3$$
$$= 323_{10}$$

Hexadecimal Code

This is a number system to the base 16 and like octal code gives a shorthand version of large binary numbers. It is particularly useful in microprocessor based systems where 8 and 16 bit binary words are used. The word hexadecimal means that six alphabetical characters are used along with those representing decimal values 0 to 9:

HEX − A − DECIMAL

 ↓   ↓   ↓

six alphabetical 0 to 9

In this way letters A to F are used to represent numbers 10 to 15 with the code shown below.

Denary	Hex.	Binary
0	0	0000
1	1	0001
2	2	0010
3	3	0011

4	4	0100
5	5	0101
6	6	0110
7	7	0111
8	8	1000
9	9	1001
10	A	1010
11	B	1011
12	C	1100
13	D	1101
14	E	1110
15	F	1111

A binary number can be represented in hex. form by breaking it down into groups of four.

Binary number = 1011 0011 1100

in hexadecimal =   B   3   C   = B3C

14_{16} = 14 to the base 16 = 0001 0100 in binary

It is much easier to express an 8 bit or 16 bit code word as a hex. number:

$E2_{16}$ = 1110 0010 as a binary word
FFF9 = 1111 1111 1111 1001 as a binary word

Self Assessment 6

1 $AD3_{16}$ — Express this hexadecimal number as a binary and as a denary value.
2 Convert the binary word 110011001 into
   (a)   octal code.
   (b)   hexadecimal code.

The need for ready made code converters is hopefully apparent now so here is a selection of common MSI types.

TTL
74184   BCD to binary
74185   Binary to BCD
7442   BCD to decimal
7443   XS3 (binary) to decimal
7444   XS3 (Gray code) to decimal.

CMOS
4028   BCD to decimal/Binary to octal.

You may not be familiar with some of these codes, they are extensively used in industry along with others, but are outside the scope of this book.

Multiplexers (MUX)

You may be familiar with the term multiplexing, and understand it to mean 'the transmission of more than one signal down a single line'. Good examples of this are frequency division multiplexing (fdm) and time division multiplexing (tdm). This definition is correct, but for our purposes it is a good idea to simplify matters and consider a multiplexer as an electronic or scanning circuit that has a number of inputs and a single output, represented by the schematic diagram of Fig. 4.22.

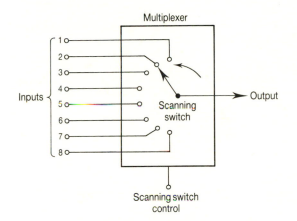

Fig. 4.22   Representation of a multiplexer

Any of the eight inputs can be connected to the output (i.e. allowed to pass through) simply by switching the scanning switch to the appropriate position. Alternatively if the switch was driven round and round, each input in turn would appear briefly at the output. This representation is purely theoretical since in practice the switching is achieved using bipolar or field effect transitors. If the scanning switch was controlled by a logic signal then the effect would be very similiar to an analogue rotary switch that was digitally controlled. Circuits like this would allow analogue signals connected to the inputs to be switched through to the output by the appropriate digital control code. Devices of this type are called Analogue multiplexers because, although the control signals are digital, the input signals could be either analogue or digital. An example of this is the CMOS 4051, eight input analogue multiplexer shown in Fig. 4.23.

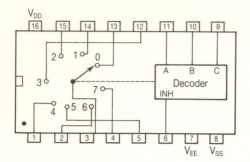

Fig. 4.23   Eight input analogue multiplexer (CMOS 4051)

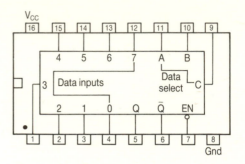

Fig. 4.24   Eight input analogue multiplexer (TTL 74151)

The operation of this integrated circuit is best explained using a table. Table 4.1

Table 4.1   Integrated circuit operation

Input channels	Channel select code		
	C	B	A
0	0	0	0
1	0	0	1
2	0	1	0
3	0	1	1
4	1	0	0
5	1	0	1
6	1	1	0
7	1	1	1

Note   the existence of an inhibit (INH) input. If a Logic 1 is applied to this there will not be a channel switched through, i.e. INH = 1 = switch open circuit.

Other examples of this type of multiplexer include:

CMOS
4051   Eight input analogue multiplexer
4052   Dual four input analogue multiplexer
4053   Triple four input analogue multiplexer

Digital multiplexers

These like the analogue devices are MSI circuits but contain only interconnected logic gates. Consequently they will only allow binary input signals to be switched or selected. Consider the eight input multiplexer shown in Fig. 4.24. The eight inputs are labelled 0 to 7 and each of these will be a binary signal. Any input can be

selected by applying the appropriate code to the data select inputs ABC. The output is obtained at Q with an inverted output provided by $\bar{Q}$. Notice the Enable (EN) input, which has a circle drawn showing that it is an active 'low', i.e. a Logic 0 must be applied to the EN input for the circuit to operate. If the EN input is take to Logic 1 it will not operate. The channel selection is achieved in the same way as for the analogue multiplexer already described.

To recap briefly: A multiplexer allows one of a number of available inputs to be selected by using a digital code word. If it is a four input multiplexer a two bit code is required, if an eight input device a three bit code etc. Now let us consider the other functions that this device can perform.

Parallel to serial converter

If a parallel binary word is applied to the inputs of a multiplexer it can be switched through by the data select signal to appear as a serial output. Consider the circuit shown in Fig. 4.25(a) of a four input multiplexer. The four inputs now hold the parallel word, each bit of which can be made to appear at the output Q by the correct data select sequence shown in the table of Fig. 4.25(b). If the parallel input word alters this change will be reflected in the serial output when the data select switching is repeated.

A slight variation on this is to use the device as a serial word generator. This is achieved by wiring the inputs permanently to Logic 1 or 0 as appropriate and so create the word. When the data select switch sequence is performed this parallel word will appear as a serial word.

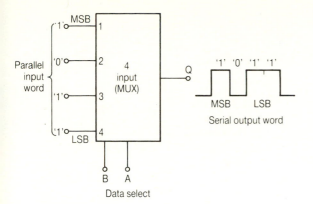

Fig. 4.25a Parallel to serial conversion using a multiplexer

Data select		Output
B	A	Q
0	0	1
0	1	0
1	0	1
1	1	1

Fig. 4.25b Data select table

THOUGHT

This sounds just like a parallel to serial converter! It is exactly the same in operation except that the input word will always be the same. The input word to a parallel to serial converter can change.

Implementation of Boolean functions using multiplexers

If you cast your mind back to the section on combinational logic you may remember that there is often a need to interconnect a number of gates so that a particular logic output function is obtained, e.g. $Q = A\bar{B}C + AB\bar{C} + \bar{A}BC$. This type of Boolean function can easily be implemented using a multiplexer. First let us draw up the truth table shown in Fig. 4.26. Notice that the

MUX inputs	C	B	A	Q	
0	0	0	0	0	
1	0	0	1	1	$A\bar{B}\bar{C}$
2	0	1	0	0	
3	0	1	1	0	
4	1	0	0	0	
5	1	0	1	1	$A\bar{B}C$
6	1	1	0	1	$\bar{A}BC$
7	1	1	1	0	

Fig. 4.26 Boolean function table

eight input permutations have each been allocated a multiplexer input (MUX input) number.

The three input conditions that produce an output of Logic 1 are indicated and all the other input conditions produce Logic 0. An eight input multiplexer can now be wired as shown in Fig. 4.27 to fulfil this requirement.

Practical Investigations 18 and 19 will give you an insight into the way these devices operate.

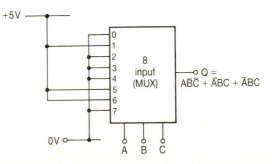

Fig. 4.27 Multiplexer implementation of a Boolean function

Demultiplexers

As the name suggests these are the inverse or opposite of the multiplexers, i.e. a demultiplexer has one input and a number of outputs as shown in Fig. 4.28. Information presented to the input is routed to its appropriate output terminal or channel. An alternative name for this device is a 'data router'.

Similar to the multiplexer, the control or data select signal determines through which channel

PRACTICAL INVESTIGATION *18*

Multiplexer operation — Parallel to serial conversion

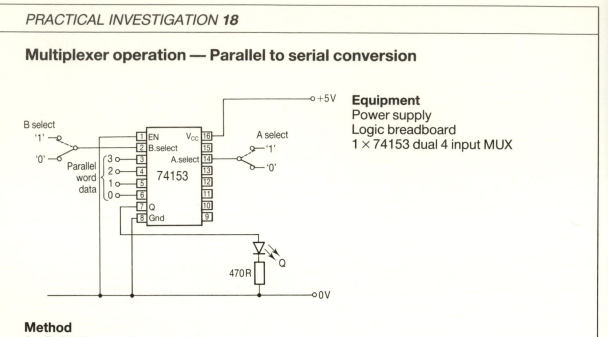

Equipment
Power supply
Logic breadboard
1 × 74153 dual 4 input MUX

Method

1 Build the circuit shown above.
2 Enter the parallel data word 1011 connecting input 0 to +5 V input 1 to 0 V inputs 2 and 3 to +5 V.
3 Operate the data select switches in the sequence shown in the truth table below, noting the resulting state of the output Q as you do so.

A	B	Q
0	0	
0	1	
1	0	
1	1	

4 Repeat the procedure indicated in step 3, but part way through the sequence change the Enable input to Logic 1 (+5 V) and note what happens.

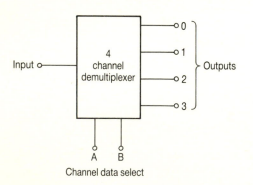

Fig. 4.28   Four channel demultiplexer

the incoming data will be routed. Many multiplexer circuits are bidirectional (particularly CMOS types) and these will function equally well as demultiplexers.

Read Only Memory devices (ROMs)

You will recall that any logic operation — e.g. adding, multiplexing, decoding, etc. — can be implemented using SSI logic gates, but for convenience it is often preferable to use commercial MSI circuits designed for the specific task. When there are many input variables even the

---

PRACTICAL INVESTIGATION *19*

Multiplexer operation — Implementing and Boolean function

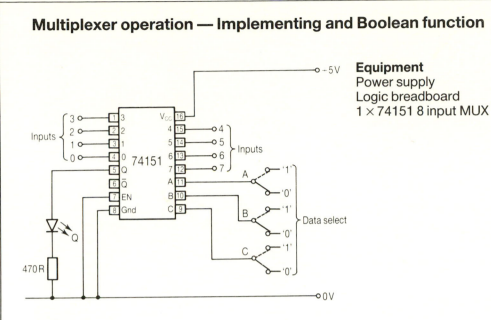

Equipment
Power supply
Logic breadboard
1 × 74151 8 input MUX

Method
1 Build the circuit shown above.
2 By wiring the inputs to +5 V or 0 V use these multiplexers to implement the Boolean function
$$Q = \bar{A}\bar{B}C + \bar{A}BC + AB\bar{C} + AB\bar{C}$$
3 Test the circuit and complete the truth table below

Input number	C	B	A	Q
0	0	0	0	
1	0	0	1	
2	0	1	0	
3	0	1	1	
4	1	0	0	
5	1	0	1	
6	1	1	0	
7	1	1	1	

4 Note the effect of taking the Enable input to Logic 1 (+5 V) at any stage in the test procedure.

use of a combination or network of MSI circuits can be cumbersome. To help solve this problem special integrated circuits are available that will perform the task of many interconnected gates.

The ROM can best be considered as an integrated circuit that has a truth table permanently stored inside it. A 16×8 bit ROM is shown schematically in Fig. 4.29.

The truth table contents are contained in 16 locations, with each location having its own special 8 bit data word written in by the manufacturer. To make any of the 16 data words

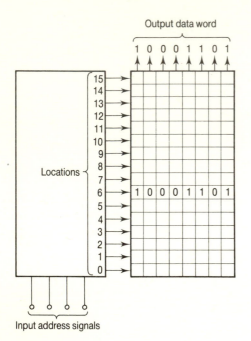

Fig. 4.29   Schematic diagram of a 16 × 8 bit ROM

appear at the output the unique address location code must be applied to the input address signal terminals. For example, suppose location six contained the word 10001101, to make this appear at the output the input signal code must be 0110. This 4 bit address code is fed into a decoder and this selects location six. In this way an 8 bit data word can be produced by a 4 bit input signal, and so this device can be used to replace complicated combinational logic circuits. ROMs tend to be available in a large format, e.g. a total of 16384 bits with 2048 locations each providing a 8 bit data word. This requires an 11 bit input address signal since $2^{11} = 2048$. Likewise a larger ROM with 4096 locations each with an 8 bit word will have a 12 bit address since $2^{12} = 4096$.

Now the problems here are:

1   The data stored in each location is permanently written in (i.e. put there) by the manufacturer during the final stages of construction using a technique called 'mask programming'. It is possible to have a ROM built to any specification, but this is uneconomical for batches of less than 100 because of the cost of setting up the equipment,

known as 'tooling up'. For this reason ROMs tend to be available to perform specific tasks notably for use in microprocessor systems and computers.

2   The information contained in each location cannot be changed or erased, hence the name Read only memory.

Programmable Logic Devices

To help overcome the problems previously outlined it is possible to obtain programmable ROMs (PROMs) where the user can program or enter the desired data into each location. This is usually a one off process so if the wrong data is programmed in — too bad! However there is also available an Erasable programmable ROM (EPROM). This allows all the data to be erased by shining an intense ultra violet light source into the window on top of the IC, Fig. 4.30. In this way the advantages of a ROM are combined with the facilities for programming each location and re-using the device at will. An important point to note here is that this facility is for *total* data erasure and not for specific locations or bits. Once programmed the window is covered with opaque tape, since normal sunlight (and other light sources) contain some ultra violet; while this will probably not erase the data it may well corrupt them.

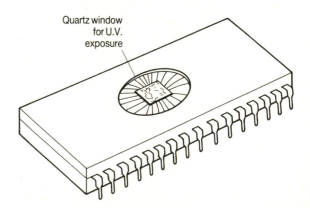

Fig. 4.30   EPROM integrated circuit

The EAROM

This recent ROM development is Electrically alterable. It performs like a standard PROM or EPROM with the advantage that while in circuit, specified data can be changed, i.e. a stored word could be altered without having to totally erase and rewrite as with the EPROM.

Programmable Array Logic and Programmable Logic Arrays

These are integrated circuits that internally consist of a matrix of AND and OR gates that are interconnected by fusible links. In the unprogrammed state all the fuses are intact so programming consists of 'blowing' the appropriate fuses to produce the required logic output. The terms programmable array logic (PAL) and programmable logic array (PLA) appear identical but there is a subtle difference, let us start by considering the PAL.

Programmable Array Logic (PAL)

Internally this can be considered as a programmable array of AND gates and a fixed array of OR gates as depicted in Fig. 4.31.

Notice that fuse links f1, f2, f4 and f7 are shown to be blown open. This gives the output $Q = A.\overline{B} + \overline{A}.B$ which is a 'sums of products' expression. AND gate arrays are frequently called 'product terms' and OR gate arrays 'sum terms'. Fig. 4.31 illustrates only a small part of the internal architecture of the device, which in practice contains a great number of gates in matrix form.

Programmable Logic Array (PLA)

Here the internal architecture allows the programming of both the AND and the OR gate arrays. This is the advantage that the PLA has when compared to the PAL, where only the AND gate array can be programmed. Fig. 4.32 shows this comparison schematically. Notice the convention that uses a cross (X) at the intersection of crossed lines to show that the fuse is intact. When programmed the diagram will indicate blown fuses by the absence of a cross at the intersection. To simplify such diagrams further only one input line is connected to a gate as shown in Fig. 4.33.

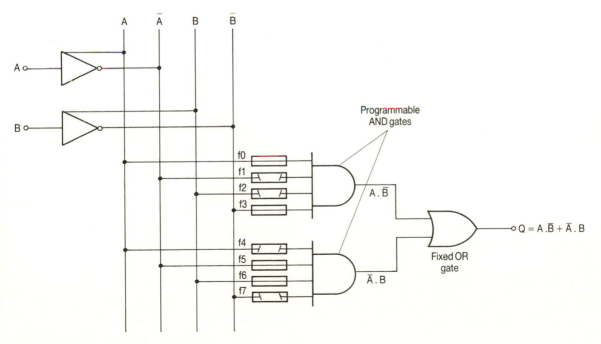

Fig. 4.31   Programmable array logic (PAL)

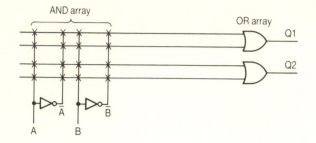

Fig. 4.32a   A 2 × 4 × 2 PAL arrangement

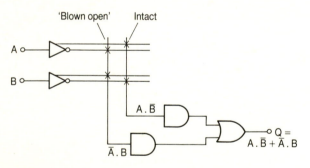

Fig. 4.32b   A 2 × 4 × 4 PLA arrangement

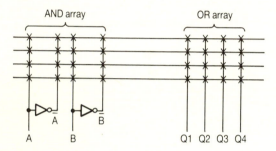

Fig. 4.33   Exclusive OR function using a PAL

THOUGHT

*This seems rather like a PROM! What is the difference
between a PROM and a PAL or PLA? A PROM usually has
a large number of inputs and address locations that can be
used to handle very complex logic functions. This
becomes very uneconomical when all the combinations
are not used. The PAL and PLA enable complex logic
functions that are below economical PROM capability to
be implemented without using SSI or MSI combinational
circuits.*

PROMs, EPROMs, PALs, PLAs require specialist equipment for programming, indeed today a number of manufacturers supply an additional software package so that personal computers can be used to simplify the task of converting the truth table into the programmed device.

Some typical programmable logic devices are:

PAL10H8 — 10 input AND/OR gate array
PAL16L8 — 16 input AND/OR invert gate array
PAL20LD — 20 input AND/OR invert gate array
PLS100N — 16 input 48 product terms — eight outputs
PLS173N — 22 input 42 product terms — ten outputs.

Sequential Logic Review

1 A bistable or flip-flop is a memory device.
2 Flip-flops can be triggered and may change state on receipt of a clock pulse.
3 Flip-flops are available as level or edge triggered devices.
4 To reduce the risk of a race hazard or glitch, edge triggered or master-slave flip-flops are used.
5 Flip-flops are manufactured to be either positive or negative edge triggered.
6 A 555 timer can be wired to provide clock pulses for sequential logic circuits.
7 Flip-flops can be interconnected to make asynchronous and synchronous counters and shift registers.
8 It is often convenient to use purpose built MSI circuits when counters or registers are required.
9 Asynchronous counters operate on a ripple through basis, with each flip-flop output providing the clock signal for the next in line. This limits the maximum operating frequency.
10 Synchronous counters use flip-flops that are all clocked together thus having a higher operating frequency than the asynchronous type.
11 A counter may be required to operate at different base numbers, e.g. binary coded decimal or octal.
12 Data can be shifted/stored or delayed using a shift register.

13 Shift registers are available that can perform the following operations:
  (a) Shift data to the left or right.
  (b) Accept serial input data and provide parallel output data.
  (c) Accept parallel input data and provide a serial output.
14 A universal shift register is a MSI circuit that will perform any shift register requirement according to how it is connected.
15 Encoder and decoder integrated circuits allow one type of binary code to be converted into another. E.g. binary, code to hexadecimal code or denary (decimal) code to binary etc.
16 A multiplexer integrated circuit can be considered as a digitally controlled sequence or scanning switch that will allow any one of a number of input signals to be connected to a single output, according to the digital control signals applied.
17 A multiplexer can also be used to provide the additional functions of:
  (a) Parallel to serial conversion.
  (b) Implementation of a Boolean function.
18 A demultiplexer is the opposite of a multiplexer, i.e. it has a single input that can be connected to any one of a number of outputs according to the control signal that is applied.
19 Many multiplexers are bidirectional, i.e. they will perform as demultiplexers.
20 A Read only memory (ROM) IC has binary information programmed in to a number of address locations. This information will appear at the output upon receipt of the correct input code.
21 ROM information cannot be changed or erased. The stored data is programmed in by the manufacturer so these devices are built to specification; these are uneconomical in batches of less than 100.
22 A PROM is a programmable read only memory device that has the data programmed in by the user — mistakes are permanent since erasure is not possible.
23 An EPROM is an erasable programmable read only memory device. This can be programmed with data that can, if required, be erased by shining an intense ultra violet light sought into the window on top of the IC.
24 An EAROM is an electrically alterable read only memory IC that can be both programmed and erased while in circuit.
25 A programmable logic device is an integrated circuit containing a matrix of interconnected gates. The links can be opened electrically allowing to create a specified output for a specified input. This allows complex logic functions to be implemented more economically than if a ROM, PROM or EPROM were used.

Self Assessment Answers

Self Assessment 6

1 $AD3_{16} = 101011010011$ binary
  $= 2771_{10}$ denary
2 The binary number 110011001 converts into:
  (a) octal code $= 631_8$
  (b) hexadecimal $= 199_{16}$.

5

Display Devices

'The visual display is the interface between us and the world of electronics.'

There is always a need to display the binary or coded output of a digital system in a form that can be understood by the operator or interested party. In the past a number of ingenious methods were developed based around the incandescent filament lamp, the neon gas indicator and the gas discharge display. Although still used today the disadvantages of these devices are:

(a) They are often large and bulky packages based around a vacuum tube.
(b) Quite high currents are involved.
(c) High voltages are required, e.g. 100–170 V for neon gas indicators and gas discharge displays.

For these reasons displays of this type tend to be limited to mains operated equipment. The display shape is determined by the shape of the filament. The neon or 'Nixie' tube display shown in Fig. 5.1(a) and 5.1 (b) is typical of this type.

Fig. 5.1a   Neon or 'Nixie' tube display

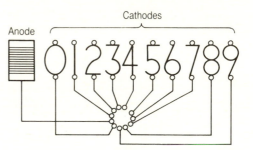

Fig. 5.1b   Neon tube pin connection

The Light Emitting Diode (LED)

If you have performed any of the Practical Investigations proposed in earlier chapters you will be familiar with the light emitting diode (LED). Essentially this is a semiconductor diode that emits visible light when it is forward biased. The intensity of the emitted light is current dependent, i.e. the higher the forward diode current (Id) the brighter the light. The maximum forward bias current (Ipk) must not be exceeded hence the need for a series current limiting resistor. Operate a LED without any regard for current limiting and it will glow very brightly for a very short time before it self destructs.

LED Construction

Light emitting diodes are constructed from a pn junction just like a normal silicon diode. Different semiconductor materials are employed in the fabrication process, usually Gallium Phosphide (GaP) or Gallium Arsenide Phosphide (GaAsP). Into this material some impurity is added, it is this together with the base semiconductor that will determine the wavelength and hence the colour of the emitted light.

The physical construction is different from that of the normal diode. The light is emitted from the junction itself, so the maximum junction area must be exposed. For the standard LED indicator a ring format is used as shown in Fig. 5.2(a).

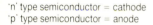
'n' type semiconductor = cathode
'p' type semiconductor = anode

Fig. 5.2a   LED pn connection

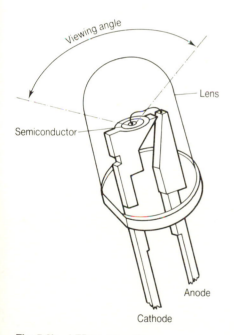

Fig. 5.2b   LED construction

The area of the diode that emits light is very small, about the size of a pin head. To improve the contrast or visibility of the device and give a wider viewing angle a coloured plastic encapsulation acts both as a filter and lens system.

This causes the emitted rays to diverge, creating the impression of a larger light source. LED indicators are available in a variety of colours, typically green, red, amber and yellow, with viewing angles up to 130°.

Driving the LED

In all the investigations in this book the integrated circuit gate has been the source of the current for the LED. Study Fig. 5.3(a). If both inputs to the AND gate are high (Logic 1) then the output Q is high and current flows from the IC through the LED which lights up. This is current sourcing therefore when Q = '1' the LED emits light.

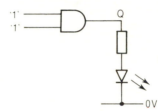

Fig. 5.3a   Current sourcing the LED

An alternative is to use the circuit of Fig. 5.3(b). Here when the output Q = Logic 0 the LED will light because the cathode is effectively taken down to the 0 V rail, and current can now flow from the +5 V supply rail through the LED to the 0 V rail. This is known as current sinking. Therefore when Q = '0' the LED emits light.

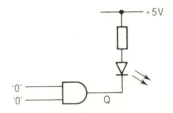

Fig. 5.3b   Current sinking the LED

The Seven Segment Display

With digital systems it is often desirable to display decimal numbers, e.g. a BCD counter counts from 0–9 in binary but a decimal display is required rather than a binary display. The plastic lens system used on LEDs determines the actual shape that the eye sees, so it is possible to use seven LEDs and arrange them in the pattern shown in Fig. 5.4.

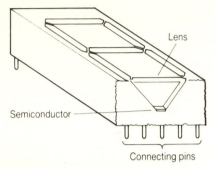

Fig. 5.4a Seven segment pattern

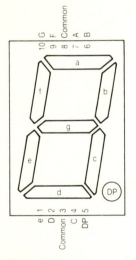

Fig. 5.4b Seven segment display construction

Each LED segment is allocated a specific letter and by energizing the appropriate segments, any number from 0 to 9 can be displayed. Thus if segments a, b, c, d and g are energized the number 3 will be shown. The physical construction of this type of device is shown in Fig. 5.4(b) and by using modern techniques such seven segment displays are obtainable from 0.3 inch up to 2.24 inches high with typical data shown in Table 5.1.

Table 5.1 LED Data

LED size (inches)	0.3	0.3	0.5	0.56	0.8	1.0	2.24
P_D max. (mW per segment)	60	75	50	81	50	89	210
Forward Current (mA) per segment (I_f) Maximum	30	30	25	60	25	25	30
Typical	10	10	20	10	10	20	20
Max. Voltage (V) $V_{R\,max}$	6.0	6.0	3.0	6.0	3.0	6.0	20.0

From this you may see that a seven segment LED display draws quite a high current. If we assume that 10 mA per segment is a typical operating current, then to display the number 8 will require 80 mA. If it is a four figure display this will require 320 mA. If such a display arrangement was used in a piece of battery operated equipment the cells would have to be physically large to cope with the demand. One method of solving this problem is to multiplex the display. In practice this means that an oscillator system controls the display energization. The required segments are illuminated or flashed in sequence very quickly so that the eye sees the complete number and not the individual flashing segments of which it is comprised. The multiplexing frequency is between 50 Hz and 250 Hz and makes a considerable current saving.

THOUGHT

Surely if the LEDs are pulsed or flashed the eye will see a continuous but dimmer display. True, to ensure that the display is bright enough the energization current is increased, this can be done without overloading the LED because it is not a continuous current. Even though increased current is used the overall power consumption is much reduced.

Practical Investigation 20 allows you to become familiar with the seven segment LED display.

If you study the data sheets on page 107 you will see that the seven segment display can be obtained in either a 'common cathode' or 'common anode' format according to the requirements.

THOUGHT

What's the difference between common anode and common cathode displays? There are seven separate light

PRACTICAL INVESTIGATION *20*

BCD Using a Seven Segment Display

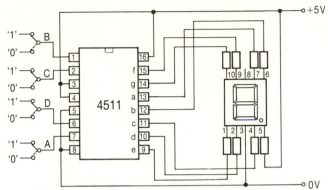

Equipment
Power supply +5 V
4511 BCD to 7 segment
Decoder/driver
Breadboard
Standard 8 segment LED display
(Common cathode)
8 × 470 R resistors

Method
1 Connect the circuit as shown above (check that your seven segment display pin connections are the same as the diagram).
2 Draw up a truth table showing how the binary input relates to the displayed number. (Input A is the least significant bit LSB.)
3 Check that the circuit does provide a display for numbers 0 to 9.
4 By investigation discover what happens when you set the inputs to numbers greater than 9.

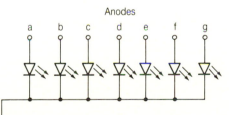

Fig. 5.5a Seven segment common cathode

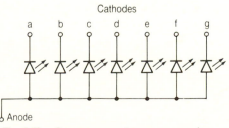

Fig. 5.5b Seven segment common anode

emitting diodes in a seven segment display. Each diode has an anode and a cathode, this makes 14 separate connections in total. By taking a common point (either the anode or cathode) the number of connections can be reduced to 8 as shown in Fig. 5.5(a) and 5.5(b).
The choice of display will be determined by:
(a) the type of logic used (positive or negative)
(b) how the display is driven, i.e. current sourcing or sinking.

In addition to the seven segment format a 'universal 1' is available. This is sometimes referred to as a '$\frac{1}{2}$ display'. It is common practice to group displays together to give the required number of digits. Fig. 5.6(a) shows a $3\frac{1}{2}$ digit group which will display a maximum of 1.999 while Fig. 5.6(b) shows a 4 digit group which will display a maximum of 9.999.

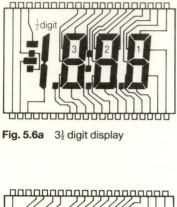

Fig. 5.6a   3½ digit display

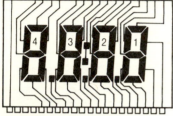

Fig. 5.6b   4 digit display

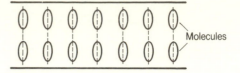

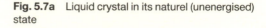

Fig. 5.7a   Liquid crystal in its naturel (unenergised) state

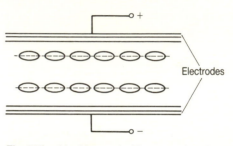

Fig. 5.7b   Liquid crystal with a potential difference applied (energised)

The Liquid Crystal Display (LCD)

This type of display now features significantly in everyday devices such as watches, clocks, calculators and the majority of hand held video games all of which use liquid crystal as the visual display medium.

The material has an organic crystalline structure that is liquid in normal use as the name suggests. When light passes through a normal liquid the molecules in the liquid cause reflection and refraction to occur. However because of the random position or orientation of these molecules the overall effect is that the light is generally scattered.

A liquid crystal has quite large rod-like molecules that exist in an ordered pattern throughout the structure of the crystal. If this pattern is altered then light passing through the crystal will be affected in a specific way. This crystal structure can be altered by the application of a voltage; consider the diagrams of Fig. 5.7(a) and 5.7(b). You can see that in the natural state the rod-like molecules have their axes in one plane throughout the material, which results in the liquid being quite transparent and light can pass through. When a potential difference is applied, as shown in Fig. 5.7(b), the molecules rotate and align with the electric field. This creates some turbulence and the light will be scattered so the crystal between the electrodes becomes opaque.

To use this effect for display purposes it is necessary to sandwich the liquid crystal between two transparent electrodes. This is achieved in practice by using two glass plates that have a thin conducting layer deposited on them of the same shape or pattern as the desired display.

THOUGHT

So it is possible to have an electrode that is transparent but still conducts electricity? Yes, usually this is in the form of a very thin layer of tin oxide. Hold a LCD watch or calculator at an angle in a strong light and you will just be able to see the seven segment electrode design on the top glass plate.

By studying the construction of the seven segment liquid crystal display shown in Fig. 5.8 you can see that where the electrodes are the transparency of the crystal can be controlled. The visibility of the display is dependent solely on ambient light, the crystal does not produce any light itself — unlike the LED.

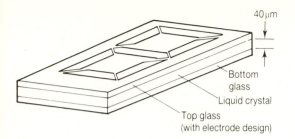

Fig. 5.8a Seven segment liquid crystal display construction

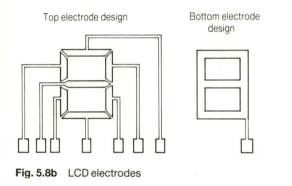

Fig. 5.8b   LCD electrodes

LCDs can be obtained that operate in transmissive or reflective mode. Let us now consider the difference.

Transmissive mode LCDs

The ambient light or light source is viewed through the crystal as shown in Fig. 5.9. Both top and bottom electrodes are transparent, with the light source either natural or artificial supplied by lamps. The energized sections of the

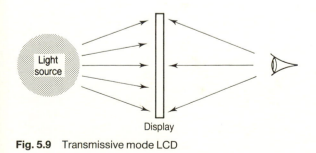

Fig. 5.9   Transmissive mode LCD

display will be visible in contrast to the transparent sections. It is now possible to build this type of display into a pilot's visor so that important information is always in the field of vision.

Reflective mode LCDs

The bottom electrode is reflecting, or a reflecting film is placed behind the bottom display glass. The illumination now comes from the same side as the observer as shown in Fig. 5.10. This type of display is commonly used for watches, calculators, clocks and instruments, since ambient lighting can be used and additional illumination provided by a lamp for poor lighting conditions.

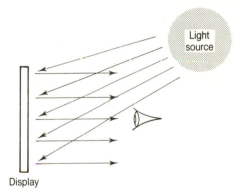

Fig. 5.10   Reflective mode LCD

The liquid crystal has a high electrical resistivity, e.g. 5 MΩ per segment of a 7 segment display. However the crystal consists of a very low compared with the equivalent LED display. However the construction consists of a liquid which is an electrolyte and two electrodes which together make up a cell. If d.c. energization is used a chemical reaction due to the electrolytic nature of the crystal will dramatically shorten the display life. Alternating current or interrupted direct current helps to increase the life of the device. For this reason liquid crystals are energized using a square wave signal of between 30 Hz and 400 Hz; this is quite easily supplied from an astable oscillator.

It is a good idea at this point to compare the liquid crystal display with the light emitting diode type, see Table 5.2.

Table 5.2 Comparison between LCD and LED display

LED	LCD
Bright display no light source required.	Display totally dependent on ambient light.
Quite high operating current required, e.g. 10 mA per segment.	Very low operating current, e.g. 120 μA all segments energized.
d.c. energization.	a.c. or square wave required (30 Hz–400 Hz) to conserve life.
Response time very fast, e.g. nanoseconds.	Slow response time, e.g. 150 ms.

From this comparison you can see that each type has its relative merits; but the LCD does have the undeniable advantage of extremely low power consumption making it very popular for battery powered equipment.

While the LED is obtainable as an individual seven segment display, the LCD is available as a package or module with $3\frac{1}{2}$, 4, $4\frac{1}{2}$, 6 and 8 digits as shown below in Fig. 5.11.

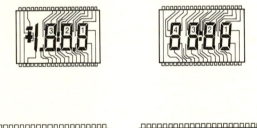

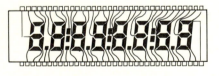

Fig. 5.11   Seven segment display packages
$3\frac{1}{2}$ digits   4 digits $4\frac{1}{2}$ digits
6 digits   8 digits

Other Types of Display

Hexadecimal Display

The seven segment, display while fulfilling the need to display numbers, does not help when letters are required. Hexadecimal code uses numbers 0 to 9 and letters ABCDEF. To display this adequately a hexadecimal Dot matrix display can be used of the type shown in Fig. 5.12.

Fig. 5.12   LED hexadecimal display

Alphanumeric Displays

These are displays that can be energized to produce numbers, letters and symbols as required. They can be the 5×7 dot matrix format of Fig. 5.13 or the starburst type of Fig. 5.14. Both of these are LED types and are available in common cathode or common anode format. In addition to this, alphanumeric dot matrix LCD modules are obtainable.

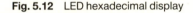

5 LED elements

7 LED elements

Fig. 5.13 5×7 dot matrix display

All the displays mentioned are TTL and CMOS compatible, but each device will require its own decoder/driver to convert the binary code into the appropriate signal for energizing the display. For this reason many displays are

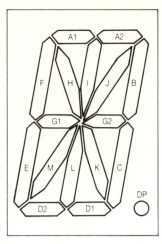

Fig. 5.14 Starburst display

obtainable with the required logic built into the display package for convenience.

Display Device Review

1 A display presents binary signals in an easily interpreted visual form.
2 Filament and neon displays were popular but tended to be bulky and had a high current or voltage requirement.
3 Light emitting diode displays are available in a range of colours and sizes.
4 A light emitting diode may be driven using current sinking or current sourcing techniques.
5 The seven segment display can be obtained in both common cathode or common anode format.
6 Current limiting resistors must be used with LEDs to prevent exceeding the maximum forward current rating (Ipk).
7 LEDs require quite a high energizing current, typically 15–20 mA per segment.
8 Multiplexing (flashing) the LED display reduces the overall power requirement of the device.
9 Liquid crystal displays (LCDs) use a crystalline medium that is naturally transparent but becomes opaque when energized.
10 A LCD consists of the liquid crystal sealed between two glass plates that each have a transparent electrode arrangement deposited on their inner surface.
11 LCDs have a very low operating current, e.g. 100 μA with all segments energized.
12 LCDs have a very short life if energized by d.c. so a square wave of between 30 Hz and 400 Hz is used for energization.
13 LCDs are available as transmissive or reflective mode devices.
    Transmissive mode: The viewer looks through the crystal at the light source.
    Reflective mode: The display is illuminated from the same side as the viewer who sees the reflected light.
14 LCDs do not generate light and require additional illumination under poor lighting conditions.
15 Hexadecimal dot matrix displays are available and will display numbers 0 to 9 and letters A to F.
16 Alphanumeric displays allow numbers, letters and symbols to be produced.
17 Dot matrix and starburst are examples of alphanumeric displays.
18 LCD alphanumeric modules are obtainable.
19 All display devices require decoder/driver circuits, and the more complicated types often have these circuits built into the display package.

6

Analogue to Digital and Digital to Analogue Conversion

'It is often required to convert a digital signal into an analogue form and vice versa.'

Digital systems and logic devices operate using some form of binary code. However many of the input signals that are fed into these circuits are in the form of an analogue voltage, i.e. a potentiometer as it moves from its minimum to maximum position may create a voltage that varies from 0 V to +10 V d.c. If this signal is to have any significance to a logic circuit it must be converted into digital form. From the other viewpoint, it may be required to take the output of a digital register containing say 0101 as a binary word and display it using a moving coil voltmeter as 5.0 V. Again conversion is required but this time the digital signal must be presented in analogue form.

A good starting point is to look at a simple analogue to digital converter and by so doing get an understanding of the prinicples involved.

The simple counter-ramp A–D converter

Study the block diagram of Fig. 6.1(a) and note the subsystems involved.

The Comparator

This is a circuit with two inputs and one output. An op-amp with an inherently high open loop gain as shown in Fig. 6.1(b) can be used as a comparator. When inputs A and B are the same

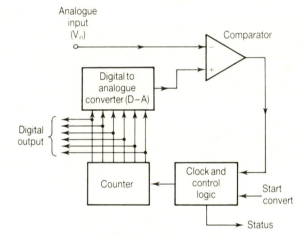

Fig. 6.1a   Counter-ramp A-D converter

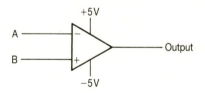

Fig. 6.1b   The op-amp comparator

the output is zero. If input A is a positive voltage greater than input B then the comparator output will be a negative voltage. If B is a positive voltage greater than A the comparator output will be a positive voltage. It is important to note here that the comparator action is 'snappy' with the output voltages clearly defined. Thus if A is slightly more positive than B the output will be

−5.0 V. As soon as B becomes slightly greater than A the output will change to +5.0 V. In this way the comparator can be used as a signalling or control device.

The clock

This is a circuit that provides rectangular switching pulses at a fixed frequency — like the one you used in the practical investigations but faster.

Control Logic

A unit that issues logic instructions to the rest of the circuit, i.e. 'start', 'stop' and 'converter busy'.

The Counter

A binary counter that increments with each clock pulse it receives.

The D–A Converter

The analogue input that is being converted into a digital representation is fed into the compara-

tor. The output of the counter must be compared with this analogue input. Since the comparator is an analogue device the digital counter output must first be coverted into an analogue signal.

THOUGHT

So this A–D converter contains a D–A converter? Yes! as indeed do many analogue to digital converters.

If you have studied the function of each block you should have a good idea of how the circuit works. Now follow the operation using the timing diagram of Fig. 6.2.

Assuming the analogue input is a positive d.c. voltage:

1 The Start convert command is issued.
2 The control circuit resets the counter to zero hence the D–A output is zero.
3 The D–A signal to the comparator is 0 V and since the analogue input is greater than this, the comparator output is −5 V (Logic 0).
4 The clock now increments the counter at a rate determined by the clock frequency.

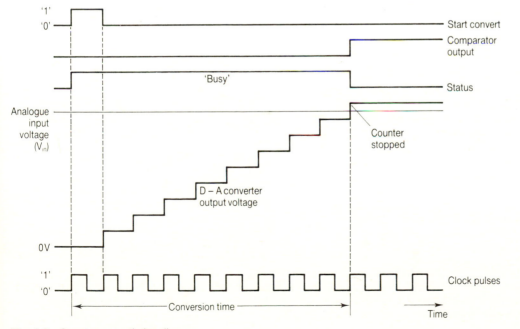

Fig. 6.2   Counter-ramp timing diagram

5 As the counter increments so the D–A output increases.
6 When the D–A output becomes slightly greater than the analogue the comparator output flips to +5 V = (Logic 1) and stops the clock.
7 The counter now holds a count which is the digital representation of the analogue input. This binary word is the digital version of the analogue input.

THOUGHT

If the comparator halts the process when the D–A output is greater than the analogue input surely an error will exist? Unfortunately this is true. A–D converters suffer from 'quantization errors'. Solutions to this problem include using very small incremental counting steps which reduce the error, and applying some 'offset' to the D–A counter so that the output binary word is either rounded up or rounded down.

The major disadvantage of this type of converter is its speed of operation. The time taken for conversion to take place will depend upon the size of the analogue input. The worst case or longest time will occur for its maximum analogue input signal. For example, a 4 bit A–D converter has a clock frequency of 50 kHz. How long will it take to convert a full-scale analogue input signal?

We know that the counter will count up to a maximum of 2^4 before it resets. Therefore the maximum count $= 2^4 = 16$. The counter increments on every clock pulse so it will take 16 clock pulses to count from 0 to 15. Each clock pulse takes $1/(50 \times 10^3)$ second $= 20$ μs. So the maximum conversion time will be

$16 \times 20 \times 10^{-6} = 320$ μs.

Self Assessment 7

1 A 10 bit A–D simple counter-ramp converter has a clock frequency of 100 kHz and a maximum analogue input of +15 V.

(a) How long will it take to convert a full-scale input signal?
(b) How long will conversion of a +6.0 V analogue input take?

The Successive Approximation A–D converter

This is a much faster device and is represented by the block diagram of Fig. 6.3. You will notice that the D–A converter, comparator, clock and control logic are present as they were in the simple ramp converter previously discussed. The difference is that the binary counter is now replaced by a binary register. Each of the bits in the register can be set and reset by the control circuits. The operation is explained in stages that can be referred to the timing diagram of Fig. 6.4.

Assume that the analogue input voltage is a steady d.c. of 13.5 V.

1 Start convert command is issued.
2 Status set to busy and first clock pulse sets register to 0000.
3 The D–A output is zero so the comparator output is −5 V (Logic 0).
4 The next clock pulse sets the most significant bit (MSB) (B4) to Logic 1.
5 B4 represents 8 V so the D–A compares this to the analogue input; since 8 V is less than 13.5 V the comparator output does not change, so B4 is kept at '1'.
6 Next significant bit is applied (B3); this represents 4 V so the output of the D–A is 12 V in total. 12 V is less than 13.5 V so B3 is kept at '1'.
7 Next bit is applied (B2), this represents 2 V making the D–A output 14 V in total. This is greater than 13.5 V so the comparator output flips to +5 V (Logic 1) and the control sets B2 to '0'.
8 Least significant bit (B1) is applied, this represents (1 V) making 13 V in total; since 13.0 V is less than 13.5 V B1 is kept at '1'.
9 All bits in the register have now been tried and conversion is complete, the status now moves to 'ready'.
10 The register word is the binary representation of the analogue input and is 1101 = 13.0 V.

You can see that there is an error present but this can be reduced by increasing the number of bits in the register. The conversion time is fast since the first pulse initializes the system and

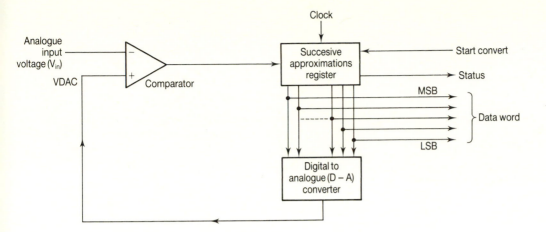

Fig. 6.3 Successive approximation A-D converter

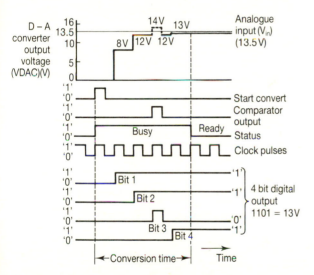

Fig. 6.4 Successive approximations timing diagram

(a) Conversion time = n + 1 clock pulses. n = 4, therefore conversion time = 5 clock pulses.
clock pulses freq = 50 kHz clock pulse period = $1/(50 \times 10^3) = 20$ μs.
Therefore conversion time = 5×20 μs = 100 μs.

(b) Since *all* bits are tried during the conversion process the conversion time is a constant, so if the analogue input is half full scale the total conversion time will be 100 μs.

Self Assessment 8

1 How fast is a 10 bit Successive Approximation A–D converter when compared to a 10 bit counter-ramp type?

subsequent pulses try the number of bits involved. This gives a total conversion time of n + 1 clock pulses, where n = number of bits in the register.

An example: a 4 bit A–D successive approximation converter has a clock frequency of 50 kHz:

(a) How long will it take to convert a full scale analogue input signal?
(b) How long will it take to convert an analogue input signal of half full scale?

Digital to Analogue Converters

The previous A–D converters used D–A converters as an intrinsic part of their function. So let us now consider how these circuits function. A D–A converter must take a binary word and provide an output that is its analogue version, i.e. 1101 = 13.0 V.

We shall start by considering a possible circuit and then look at a very common type.

The Weighted Resistor D–A converter

Consider the circuit of Fig. 6.5, which shows an operational amplifier connected as a summing amplifier. (The operational amplifier is covered fully in the companion volume: *Analogue Electronics*, Morris 1991.)

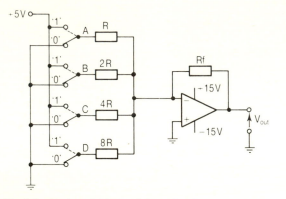

Fig. 6.5   Basic weighted resistor network

Note   that this is an inverting Amplifier.

The voltage gain $Av = \dfrac{-R_f}{R_{in}}$ since $A_V = \dfrac{V_{out}}{V_{in}}$.

$$-V_{out} = \frac{R_f}{R_{in}} \times V_{in}.$$

Now suppose $\dfrac{R_f}{R} \times 5.0 \text{ V} = 8.0 \text{ V}.$

When switch A is in the +5 V (Logic 1) position $V_{out} = 8.0$ V. When switch B is in the +5 V position (Logic 1) V_{out} will be

$$\frac{-R_f}{2R} \times 5.0 \text{ V} = -4.0 \text{ V}$$

Likewise switch C at Logic 1 will give -2.0 V out, and switch D at Logic 1 will give -1.0 V out.

You can see now that A is the MSB and D the LSB, so if the input binary word to this circuit was 1011 the output would be

$$8 \text{ V} + 0 \text{ V} + 2 \text{ V} + 1 \text{ V} = -11 \text{ V}.$$

This is a perfect analogue representation of the input binary word.

This circuit uses the values or weightings of the input resistors to achieve the required binary progression between the bits. This appears

to be a good idea in theory but in practice presents problems:

1   The weighting must follow the binary progression exactly, i.e. R, 2R, 4R, 8R, 16R, etc.
2   The ratio between each weighted resistor and R_f determines V_{out}.

The practical circuit of Fig. 6.6 illustrates these problems. You can see that +5 V on the MSB

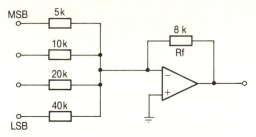

Fig. 6.6   Practical weighted resistor network

will give $V_{out} = (-8 \text{ k}/5 \text{ k}) \times 5.0 \text{ V} = -8.0 \text{ V}$ and that +5 V on the LSB gives $V_{out} = (-8 \text{ k}/40 \text{ k}) \times 5.0$ V $= -1.0$ V. However you may notice that the resistor values have to be exact and are not all going to be preferred values. Also the tolerances of the individual input resistors and R_f will combine to cause errors. The problem becomes greater when you start to consider 8, 10, and 12 bit converter circuits. These would only be possible using custom made resistors. Even if the tolerances were the same, since resistors of many megohms down to several

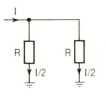

Fig. 6.7a   Parallel resistor network

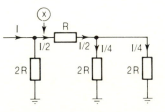

Fig. 6.7b   Series/parallel resistor network

kilohms would be involved, errors would still exist.

The R–2R Ladder Network

The problems that existed with the weighted resistor network are solved using this arrangement. Let us start with a basic arrangement and work up to the full R–2R ladder network. Fig. 6.7(a) shows two resistors connected in parallel. Since both resistors are of the same value the current (I) splits when it reaches the node and the same current will flow in each resistor namely I/2 (or $\frac{1}{2}$I if you prefer). A development of this is shown in Fig. 6.7(b).

   Now the resistance placed to the left of point x is 2R. The resistance to the right of x is given by:

R in series with 2R in parallel with 2R
or   Req = R + 2R‖2R   where   Req = equivalent resistance and ‖ means in parallel with, this mean that Req = 2R.

So the resistance to the right and left of x = 2R. The current will split as shown with I/2 flowing through R and I/4 flowing through the parallel 2R resistors. Notice how we are starting to get natural binary progression in terms of the circuit currents I, I/2 and I/4. This circuit can be extended to give a ladder network as shown in Fig. 6.7(c).

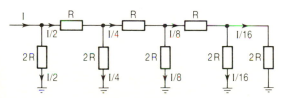

Fig. 6.7c   R-2R ladder network

You may now see how the current divides in the circuit to give the binary progression in each of the 2R resistors. This means that the resistance to the right and left of each node is 2R — if you doubt this then work it out.

THOUGHT

So the resistance when 'looking into' the R–2R ladder is 2R? CAREFUL! If you work from the back towards the front you will find that you finish up with 2R in parallel with 2R making R. The input resistance to the R–2R ladder network is R.

The significance of this circuit is that however many bits are involved the only resistors required have the values R and 2R. These can readily be fabricated in integrated circuit form so that the entire ladder is formed using resistors that have equal stability and tolerance values. This is achieved by using a single slice of semiconductor for the fabrication process. The resulting resistors will all have similar physical characteristics.

   A typical 4 bit R–2R ladder D–A converter is shown in Fig. 6.8. Note that the voltage source for the ladder network is a stable reference voltage V_{ref}. The output voltage (V_{out}) of the operational amplifier is given by

$V_{out} = I_{in} \times R_f$ where I_{in} = input current from the ladder, and R_f = feedback resistance.

The 4 bit register holds the data word that is being converted, e.g. 1011, and these logic outputs control electronic switches. So for the example quoted the currents I/2, I/8, and I/16 will flow because these switches are in the Logic 1 position while the I/4 switch is at Logic 0.

   The input to the op-amp will be given by:

$I_{in} = I/2 + I/8 + I/16$

for the circuit shown in Fig. 6.8

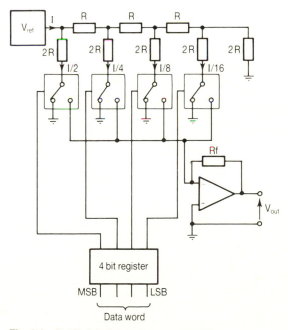

Fig. 6.8   R-2R digital to analogue converter

PRACTICAL INVESTIGATION *21*

The R–2R Ladder D–A Converter

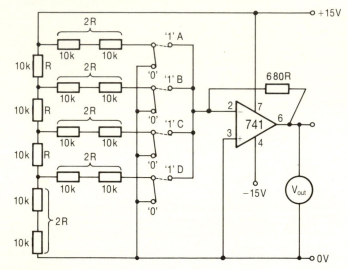

Equipment
Dual power supply (±15 V)
0–15 V voltmeter
Breadboard
741 operational amplifier
13 × 10 k resistors
680 R resistor

Method
1   Construct the circuit shown above and monitor V_{out} with the voltmeter.
2   Draw up a four variable input truth table in the form shown giving all the possible binary inputs 0000 to 1010 (1 to 10) together with the analogue output voltage, V_{out}.

A	B	C	D	V_{out}
0	0	0	0	
0	0	0	1	
0	0	1	0	
0	0	1	1	
0	1	0	0	

3   By investigation enter the output voltage for each of the binary inputs.

Results
1   What are the incremental steps in V_{out}?

$V_{ref} = +5$ V
$R = 10$ k0
$R_f = 32$ k0

Calculate the output voltage V_{out} for the data word 1001.

Method
1   Calculate I.

$$I = \frac{V_{ref}}{R}$$

(remember the input resistance to a R–2R ladder is R)

$$I = \frac{5.0}{10 \times 10^3} = 0.5 \text{ mA}$$

2   Calculate I_{in}.

$I_{in} = I/2 + I/16 = 0.25$ mA $+ 31.2$ μA
   $= 0.281$ mA.

3   Calculate V_{out}.

$V_{out} = R_f \times I_{in} = 32k \times 0.281 \, mA = 8.99 \, V$
(9 V).

It is worth noting that the resistors R in conjunction with R_f will determine the actual output voltage, and R_f can be chosen to provide the desired full scale output. By working through Practical Investigation 21 you can explore the R–2R D–A converter further.

Purpose built converters

As with many of the circuits considered in this book it is possible to obtain A–D and D–A converters in a single IC package. The data sheets on pages 118–119 show a selection from the vast choice that is available. We have considered only the basic counter ramp and successive approximation types, but very much faster 'flash' converters are available that use different techniques to obtain very short conversion times. Each converter offers advantages that make it suitable for a particular application.

Selection of the appropriate type is achieved by choosing one that matches the required specification. Interestingly there are some dual function converters, one such device is the ZN425E which we shall now study in some detail.

The ZN425E 8 bit D–A/A–D converter

This is a 16 pin dil IC that will function as an 8 bit digital to analogue *or* analogue to digital converter according to how it is connected. Fig. 6.9(a) shows the pin connections and Fig. 6.9(b)

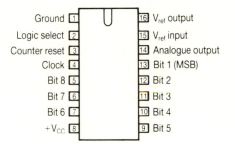

Fig. 6.9a   ZN425E pin connections

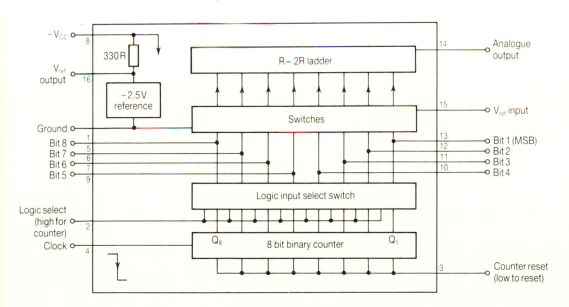

Fig. 6.9b   ZN425E block diagram

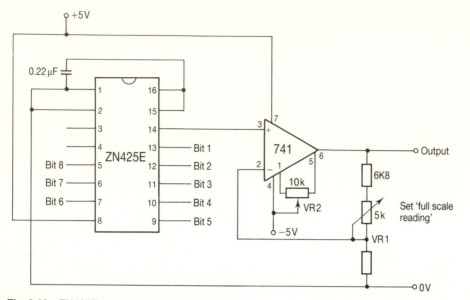

Fig. 6.10   ZN425E as an 8 bit D-A converter

is a block diagram showing the main internal circuits.

Notice that this integrated circuit contains:

An 8 bit R–2R ladder D–A converter
A 2.5 V precision voltage stabilizer
An 8 bit binary counter
8 binary switches
Switch control logic.

There is also a facility for using an external reference voltage (V_{ref} input) should this be preferred.

This collection of circuits within one package means that relatively few external components are required to create the converter of your choice. Consider Fig. 6.10, which shows the ZN425E connected as an 8 bit D–A converter. The output from pin 14 is already an analogue voltage, so a further amplifier is not required to convert the R–2R ladder current into a voltage. The operational amplifier shown in the circuit is acting as a buffer, with a variable resistor (VR1) that allows the full scale output to be adjusted for calibration, and preset (VR2) which allows d.c. offset to be removed.

A–D Conversion using the ZN425E

The circuit of Fig. 6.11 shows the ZN425E connected as an A–D converter. You may notice that it is a slightly more complex circuit since a comparator (531) and control logic (7400) must be added, together with clock pulse and command signals. This circuit is a counter type converter that operates in the following way.

1   The negative edge of the convert command pulse resets the counter (to zero) and the status sets to high (Logic 1).
2   The positive edge of the convert command lets the clock start incrementing the counter.
3   When the analogue output from the R–2R ladder D–A circuit equals the analogue input, the comparator inhibits the clock pulses, i.e. stops them activating the counter.
4   The counter is now stopped and the status is now low (Logic 0) indicating that the 8 bit digital output represents the analogue input.

The conversion time is dependent upon:

(a)   The analogue input.
(b)   The clock frequency.

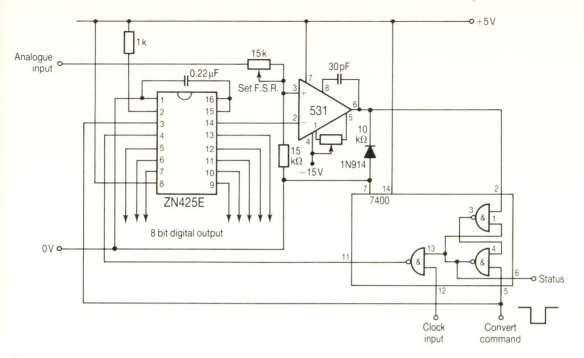

Fig. 6.11   ZN425E as an 8 bit A-D converter

Conversion time for a full scale input will be:

$2^8 \times$ clock pulse periodic time.

Sample and Hold circuits

Any analogue to digital converter (even the fast 'flash' converters) will take a finite time to complete the conversion process. During this time the analogue voltage must be held constant or the converter will be 'chasing', trying to convert an ever changing signal. Consider the waveform shown in Fig. 6.12. If the start convert command is issued at time t1 then somehow this voltage level must be held until the conversion is complete, yet you can see from the waveform that it will be increasing during the conversion time. The solution to this problem is to use a sample and hold circuit. This is basically a storage capacitor and a switch, represented diagrammatically in Fig. 6.13.

If the switch is closed at sample time t1 the capacitor will charge to 3.0 V. If the switch is now opened, the 3.0 V charge is held in the

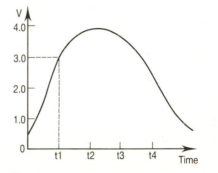

Fig. 6.12   Analogue input voltage waveform

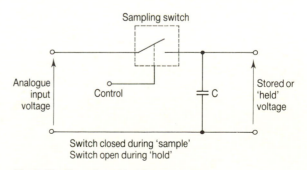

Switch closed during 'sample'
Switch open during 'hold'

Fig. 6.13   Sample and hold arrangement

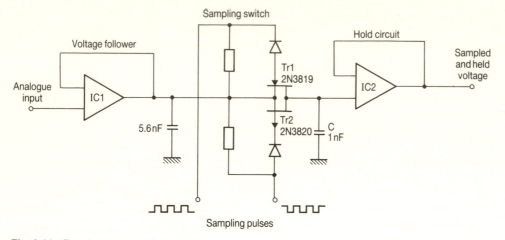

Fig. 6.14   Practical sample and hold circuit

capacitor (C) and can be converted by the A–D converter. While in Hold mode the analogue input can vary without affecting the voltage that is being converted. There are however two points that must be considered with regard to this type of circuit:

1   There must be a facility that enables C to discharge as well as charge, so that the charge may follow the analogue voltage fluctuations exactly.
2   Capacitor C must feed into a high impedance circuit otherwise the charge will be dissipated.

Figure 6.14 shows how these factors are dealt with in a practical circuit. The two amplifiers have unity gain and act as buffers, each having a very high input impedance and low output impedance. The electronic switch consists of two junction gate field effect transitors. Tr1 is an 'n channel' Jfet and Tr2 a 'p channel' Jfet. These are *both* turned on by a pair of control pulses (note that the pulse for Tr2 is in anti-phase to that for Tr1). This means that both transistors are 'on' together. Tr1 is available to pass current from IC1 into C, raising the charge if the analogue input is greater than the stored voltage; but if the analogue voltage is lower than that sorted in C, Tr2 will allow current to pass from C back into the output impedance of IC1 thus lowering the stored charge. In this way the charge stored in C follows variations in the analogue signal during the sample time. The

very high input impedance of IC2 prevents C from discharging too much during the hold time. The waveform diagram of Fig. 6.15 illustrates the important characteristics of a sample and hold circuit.

The *acquisition time* is the time taken for the capacitor to charge to the analogue voltage after the sample instruction has been issued. This is defined as the time taken for C to charge to within a percentage of the maximum, e.g. 3 ns to 0.01%.

The *Aperture time* is the time between the hold instruction being issued and the electronic switches actually opening.

Droop refers to the loss of charge from C and consequent reduction in Vc after the switches have opened. It is quoted in terms of volts per second, i.e. 1 mV/s. Ideally a short acquisition and aperture time are desired with zero droop occurring. A low value C gives a short acquisi-

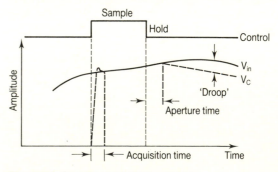

Fig. 6.15   Sample and hold waveforms

tion time but poor droop, while a large value C gives less droop but a longer acquisition time.

Sample and hold circuits are available in integrated circuit form. A precision type AD585AQ is shown in Fig. 6.16.

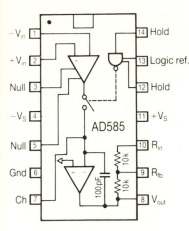

Fig. 6.16 Precision sample and hold circuit

Converter Review

1 A simple counter-ramp converter converts an analogue voltage into a digital data word, using a clock, counter, comparator and D–A converter.

2 For a counter ramp converter maximum conversion time is given by:
conversion time = $2^n \times$ clock pulse period
(where n = number of bits).

3 A successive approximation converter always takes the same time to convert any analogue signal since:
conversion time = $(n + 1) \times$ clock pulse period
(where n = number of bits).

4 A digital to analogue converter converts a digital data word into an analogue voltage.

5 A weighted resistor D–A converter uses precision resistors and is difficult to construct beyond 4 bits.

6 R–2R D–A converters are popular because they use only two values of resistor regardless of the number of bits involved.

7 D–A and A–D converters are available in integrated circuit form.

8 A sample and hold circuit must be used in conjunction with an A–D converter in order to keep the analogue signal constant during conversion.

9 Sample and hold circuits consist of an electronic switch and a storage capacitor.

10 The acquisition time is the time a sample and hold capacitor takes to charge to within a quoted percentage of the maximum input, e.g. 4 ns to 0.01%.

11 The aperture time is the time the electronic switch takes to open after the hold instruction has been issued, e.g. 4 ns.

12 Droop rate refers to the loss of charge from the storage capacitor quoted as volt drop per second, e.g. 2 mV/s.

13 Sample and hold circuits are available in integrated circuit form.

Self Assessment Answers

Self Assessment 7

1 (a) Full scale conversion = $2^{10} \times$ clock period

$$\text{clock period} = \frac{1}{100 \times 10^3} = 10 \ \mu s$$

$2^{10} = 1024$ pulses

Therefore conversion
time = $1024 \times 10 \times 10^{-6} = 10.24$ ms.

(b) Full scale input = +15 V
to convert this takes 1024 pulses

$$\text{Volts per pulse} = \frac{15}{1024} = 14.65 \ \text{mV/pulse}$$

No. of pulses required to convert +6.0 V

$$= \frac{6.0}{14.65 \times 10^{-3}} = 409.5 \ \text{pulses}$$

each pulse takes 10 μs.

Therefore conversion time for a 6.0 V analogue input is:

$409.55 \times 10 \times 10^{-6} = 4.10$ ms.

Self Assessment 8

---

1 10 bit successive approximation type takes

10 + 1 clock pulses = 11

10 bit counter ramp type takes 2^{10} clock
pulses = 1024
Therefore the successive approximation type
is:

$$\frac{1024}{11} = 93 \text{ times faster.}$$

7

Fault Diagnosis

When building and testing the practical investigations in this book you probably found that the circuit did not always work first time, usually because of a construction error. This is very demoralizing, in fact the disappointment and frustration is often so great at this point that many people give up and discard the circuit.

If you persevere however and check your circuit carefully against the circuit diagram, ensuring that all links are in place and connections made then it should work! If all is still not well, the reason may be:

(a)   The circuit diagram is incorrect.
(b)   The integrated circuit (chip) is faulty.
(c)   An external component is faulty.

Let us assume that the diagram is correct, this limits the field to either a discrete component or integrated circuit fault.

Hold it!   Before launching into a frantic search for the rogue component there are a number of other things that must be checked because they are so obvious.

1   Check that the power supply is connected to the circuit.
2   Check that the power supply is on and delivering the appropriate voltage.
3   Check that the power supply connections to the individual integrated circuits have been made; it is surprisingly easy to overlook these when building a circuit.

Discrete components tend to fail by developing an open or short circuit or undergoing a resistance change. The methods of diagnosing component faults in analogue equipment is covered in the companion volume to this (*Analogue Electronics*, Morris 1991). Briefly the common methods consist of tracing a signal through the circuit and measuring voltage levels at various points to identify the faulty component.

Digital circuits are somewhat different. Few discrete components are used and they are external to the main logic circuitry which is in an encapsulated package. It is quite likely that when a logic circuit fails the chip may be faulty. Integrated circuits are very reliable but they can go wrong, particularly if they are abused (intentionally or otherwise) in the following ways:

1   Pins shorted together (possibly with excessive solder).
2   Connecting the power supply to the wrong pins.
3   Using power supply voltages that are higher than those specified.
4   Subjecting the device to excessive heat, i.e. while soldering or desoldering.
5   Not protecting CMOS devices from electrostatic charge when building the circuit.

Remember   If you have wired a circuit incorrectly and found that it does not work, this brief test may have damaged the IC! When an IC has failed it is the job of the electronic technician to diagnose the fault, replace the component and test the equipment, recalibrating it if necessary. Let us initially start by listing the types of fault that are likely to occur in a digital circuit:

1   IC inputs open or short circuited.
2   IC outputs open or short circuited.
3   IC inputs or outputs shorted to ground (0 V rail).
4   IC output permanently low (Logic 0).
5   Internal IC fault leading to wrong sequencing, i.e. incorrect counting or shifting.

You have probably built your circuit on a

breadboard which is very convenient for assembly, testing and the recovery of components. Commercially built circuits however are constructed using printed circuit board (pcb) techniques. This allows a large number of components to be grouped into a physically small space on one side of a board, with all the interconnections made using thin copper tracks on the opposite side of the board. Construction methods like this enable complex electronic circuits to be built so that they occupy the minimum of space. This desire for minimization has led to the development of single ICs that contain vast amounts of circuitry (VLSI), and other improvements like surface mounted devices and double sided pcbs that enable further cramming. While this leads to a desirably compact world it does complicate things for the service engineer. Cost effectiveness demands that the process of fault diagnosis be carried out speedily and efficiently; unfortunately the dismantling of the equipment and replacement of a faulty component is often extremely time consuming and difficult, because of the assembly methods used to achieve the sleek compact appearance. You may have experienced a similar circumstance if an instrument indicator lamp has failed on your car dashboard. You know what has gone wrong, you know where it is, you have the correct spare part and it will take 10 seconds to fit. However such is the design of the car that to get to the lampholder the dashboard must come out; to do that the parcel shelf and the steering wheel must be removed, the speedometer must be disconnected and then there are only 24 screws to remove and so on. Manufacturers with good forethought always design their equipment so that the servicing and maintenance can be achieved with the minimum of unnecessary dismantling — may all the equipment you work on be of this type!

Now to the business of fault diagnosis on digital circuits. The usual analogue equipment i.e. signal generator, oscilloscope and voltmeter, is not going to be of much help to us when dealing with logic. The voltmeter will only be of use for checking the power supply levels. The oscilloscope cannot usefully be employed unless the signals are repetitive as in a clock pulse generator. Any signals that may be required will be binary and must be compatible to the type of logic employed, i.e. TTL or CMOS. The equipment required to help diagnose faults in digital circuits is specialized so now is the time to identify the common types of equipment and see how they can be employed.

The Logic Probe

This is shown in Fig. 7.1. It is basically a slender

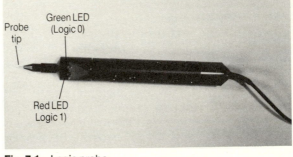

Fig. 7.1   Logic probe
[RS catalogue, Nov. 1990–Feb. 1991, p. 1093]

probe with two LEDs near the tip. This device derives its power from the circuit under test (usually via the crocodile clips connected to the supply rails). The green LED indicates Logic 0 (Low) and the red LED Logic 1 (High). There are a variety of probe types available but each performs the logic function outlined above, with variations such as TTL or CMOS operation or a built in memory that allows very fast pulses to be displayed.

THOUGHT _____

What is the difference between a logic probe and a single LED? The probe is designed to display the logic levels. This means it has switching thresholds. If you cast your mind back to Chapter 2, TTL Logic 0 = 0 V to 0.8 V. While Logic 1 = 2.0 V to 5.0 V.

_____

An LED will glow at voltages about about 1.0 V. The logic probe however would behave thus:

Voltage range 0 V to 0.8 V = Green ('0')
Voltage range 2.0 V to 5.0 V = Red ('1')

As with all test instruments this display must be interpreted, and to do this the possible LED permutations must be considered.

Red on
Green off   Logic 1

Red off
Green on   Logic 0

Red off
Green off   No connection

Both flashing = pulse waveform below 100 Hz
Both on = pulse waveform above 100 Hz.

When both LEDs are illuminated the brightness of green compared to red will give a rough indication of the mark to space ratio of the pulse waveform.

A logic probe can be used to check the logic state of each pin of an integrated circuit. This can help to find a faulty IC in a digital circuit and determine which parts of the circuit are functioning correctly.

Logic clip or monitor

This is a spring loaded device that clips directly on to the top of an integrated circuit as shown in Fig. 7.2. There is an LED display for each pin. It is really like a multiple logic probe in that the logic state for each pin is displayed simultaneously. LED 'on' = Logic 1 (High), LED 'off' = Logic 0 (Low). *Note* that unconnected inputs to an integrated circuit usually float high. This is particularly useful for checking the operation of counters, flip-flops and shift registers. These devices are usually available to fit 16 pin integrated circuits but they will also fit 14 and 8 pin circuits.

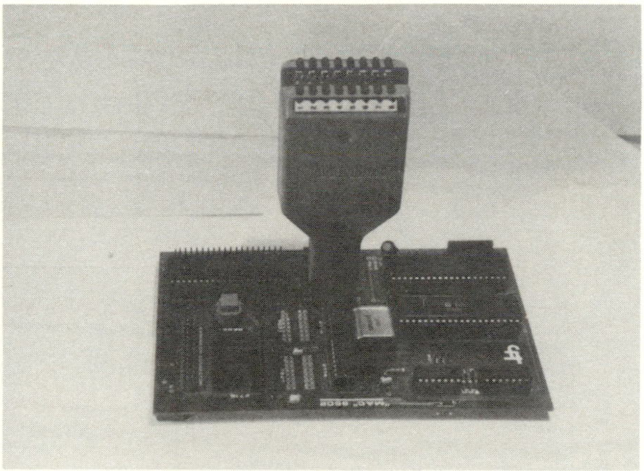

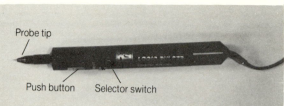

Fig. 7.2   Logic clip (monitor)

Logic pulser

This probe shown in Fig. 7.3 is of similar design

Probe tip

Push button   Selector switch

Fig. 7.3   Logic pulser

to the logic probe except that it is a device for injecting binary pulses into a logic circuit for test purposes. The pulse derives its power from the circuit under test (via crocodile clips); when the button is pressed a single binary pulse is delivered. Some pulses have a switch that is marked:

1, 4, Cont, where:
Position 1   = 1 pulse when button is pressed
         4   = 4 pulses when button is pressed
      Cont = A continuous train of binary pulses is delivered for as long as the button is pressed.

Interestingly, pulsers are designed with tri-state outputs, which means when not actually delivering a pulse the output is a high impedance that will not disturb the logic circuit to which it is applied. This is a useful test device for supplying pulses to counters and registers to check their operation.

When used in conjunction with the logic probe they can be used to test individual integrated circuits and diagnose open and short circuits.

Practical Investigation 22 will familiarize you with the application of the probe and pulser.

When using the pulser and probe you should have noted that where a short circuit existed (either to the supply or 0 V rail) the probe LED did not indicate a change of state, even when a pulse was applied. Where an input to the gate was floating or disconnected the pulser and probe could be used to prove that the gate itself was functioning correctly.

It is interesting to note that the logic probe

PRACTICAL INVESTIGATION *22*

Use of the Logic probe and pulser

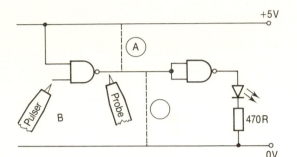

Equipment
Logic probe (CMOS or TTL)
Logic pulser (CMOS or TTL)
NAND gate (4011 or 7400)
power supply
Breadboard

Method
1 Build the circuit shown above noting that one input to the NAND gate is left floating, i.e. unconnected.
2 Using the logic probe check the state of the inputs and outputs of each gate.
3 Using the probe to monitor the output of the NAND gate, apply a pulse to the floating input and observe. The probe light should blink momentarily — this shows that the pulse applied to the NAND input has caused the gate to switch, therefore the NAND gate is satisfactory.
4 Monitor the NOT gate output with the probe and apply a pulse to the floating input.
5 Repeat the above procedures (3) and (4) when

    (a) The NAND gate output is shorted to the supply rail (wire links inserted as indicated by the dotted line A).
    (b) The NAND gate output is shorted to the 0 V rail (wire links inserted as indicated by the dotted line B).

Results
    Make an observation about the types of fault that were introduced in this circuit and how they were indicated by the probe.

will indicate changes of state that are not readily displayed by the indicator LED.

Instruments for measuring current, voltage and resistance in digital circuits

Individual ICs tend to be low-power devices, consequently logic circuits can be built using thin wires or copper tracks in the knowledge that the current flowing will be very low and the

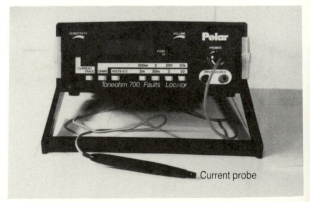
Current probe

Fig. 7.4 Fault locator

PRACTICAL INVESTIGATION *23*

Current Tracing

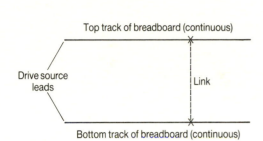

Top track of breadboard (continuous)

Drive source leads

Link

Bottom track of breadboard (continuous)

Equipment
Breadboard
Current tracer
Wire links

Method
1 Connect the instrument drive leads to the breadboard as shown above and place the link shorting the two tracks together.
2 Holding the current sensing probe at right angles to one of the drive leads adjust the instrument sensitivity until an audible output is heard with the probe about 5 mm from the wire.
3 Now trace the current along the entire length of the upper and lower tracks.
4 Observe the effect of turning the probe tip so that it is parallel to the track.
5 Trace the current through the short circuit link.
6 Repeat steps (4) and (5) for different sensitivity settings.

Results
This investigation shows how a current tracer can be used to pinpoint printed circuit board track faults without the need for dismantling.

---

voltages employed will also be low. This set of conditions has led to the development of special test equipment monitoring very low currents, voltages and resistance. The facilities are usually available in one instrument called a 'Fault locator'. Fig. 7.4 shows a typical instrument of this type that offers the following facilities:

Current tracing
Milliohm measurement
Millivolt measurement.

Current Tracing

Commercial logic circuits are built on printed circuit board. The tracks are often very thin and very close together. There is always the possibility that a short circuit will occur between tracks,

caused by a sliver of solder, wire or dirt, etc. providing a conducting path. When the circuit operates, current will flow down the intended tracks and be redirected via the short circuit. A current tracer exploits the principles of electromagnetic induction to sense the current flowing in the copper tracks as shown in Fig. 7.5.

Inside the tip of the probe is a transducer that will detect a small magnetic field (Hall effect

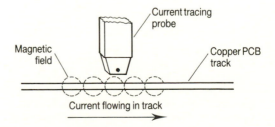

Current tracing probe

Magnetic field

Copper PCB track

Current flowing in track

Fig. 7.5 Current tracing principle

PRACTICAL INVESTIGATION *24*

Short Circuit Location using a milliohm meter

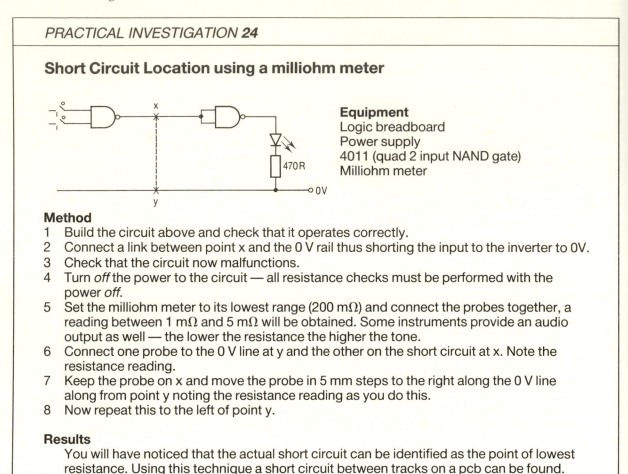

Equipment
Logic breadboard
Power supply
4011 (quad 2 input NAND gate)
Milliohm meter

Method
1 Build the circuit above and check that it operates correctly.
2 Connect a link between point x and the 0 V rail thus shorting the input to the inverter to 0V.
3 Check that the circuit now malfunctions.
4 Turn *off* the power to the circuit — all resistance checks must be performed with the power *off*.
5 Set the milliohm meter to its lowest range (200 mΩ) and connect the probes together, a reading between 1 mΩ and 5 mΩ will be obtained. Some instruments provide an audio output as well — the lower the resistance the higher the tone.
6 Connect one probe to the 0 V line at y and the other on the short circuit at x. Note the resistance reading.
7 Keep the probe on x and move the probe in 5 mm steps to the right along the 0 V line along from point y noting the resistance reading as you do this.
8 Now repeat this to the left of point y.

Results
You will have noticed that the actual short circuit can be identified as the point of lowest resistance. Using this technique a short circuit between tracks on a pcb can be found.

---

device or current transformer). Low direct currents are difficult to detect so it is usual for the current tracer to work in conjunction with a pulsed current source, supplied either by the instrument itself or the logic pulser mentioned previously. During current tracing operations the logic circuit is not powered, the only current that flows is the pulsed drive current for the purposes of current tracing. This voltage must be of a low amplitude so as not to damage the integrated circuits, typically 0.5 V at 50 kHz.

The instrument will indicate current flow by producing an audio output, meter reading, or causing a light to glow. In this way the current flow can be monitored by noting the audio tone, meter reading or light intensity. By tracing the current along the PCB tracks the short circuit

can be located. Consider Fig. 7.6.

The short circuit diverts the current flow through the circuit and can be detected and traced using the probe.

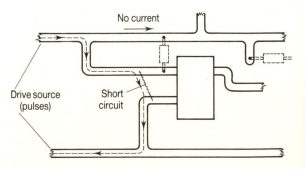

Fig. 7.6 Current tracing

PRACTICAL INVESTIGATION *25*

Fault Location using a mV Meter

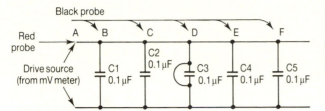

Equipment
Breadboard
$5 \times 0.1 \, \mu F$ capacitors
Millivoltmeter
Wire links

Method
1 Build the circuit shown above, placing a link across C3 to simulate a short circuit.
2 Ensure the drive source leads are connected to the circuit and attach the red voltmeter probe to point A.
3 Using the other probe measure the voltages at points B, C, D, E, F, and enter the results in a table similar to that shown below.

Test points	A–B	A–D	A–C	A–E	A–F
Voltage					

4 Repeat steps (2) and (3) with the link positioned
   (a)   across C4
   (b)   across C5.

Results
You should have found that the measured voltage drop increased with the distance from point A, but beyond the short circuit all the voltages were the same. Using this method a short circuit component on a printed circuit board track can be located.

Note Current tracing cannot be performed on circuits that have capacitors connected between the tracks. The pulsed drive source will cause the capacitively coupled tracks to appear as short circuits.

Practical Investigation 23 is relatively simple and will help you to get the feel of current tracing.

Using the Milliohm Meter

Where a short circuit exists there must be a low resistance. This low resistance will be a fraction of an ohm (milliohm). Resistance measurements are made under d.c. conditions and so capaci-

tors between the tracks do not present a problem.

Consider the diagram of Fig. 7.7. The resist-

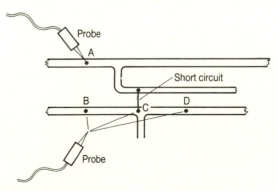

Fig. 7.7 Resistance check for a short circuit

PRACTICAL INVESTIGATION *26*

Fault Diagnosis Exercise

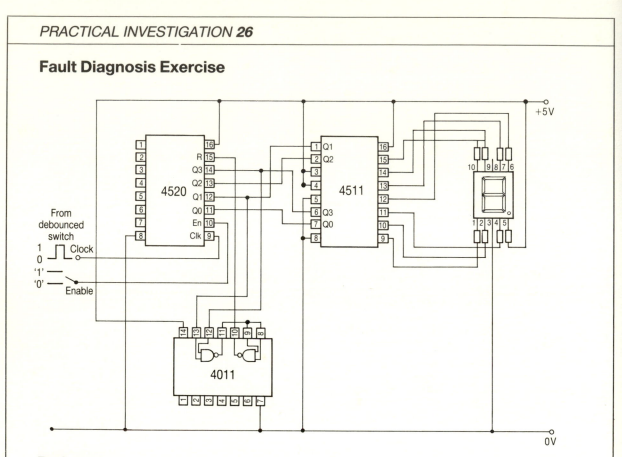

Equipment
Breadboard
Debounced switch
Two switches
Seven segment display
8 × 470 R resistors
4520 dual 4 bit binary counter
4511 7 segment decoder/driver
4011 quad 2 input NAND gate
Power supply
Logic probe
Logic pulser
Logic monitor

Method
1   Build the circuit shown above.
2   Set the enable switch to Logic 1 and test the circuit using the debounced switch to
    supply clock pulses. The circuit should count and display pulses 0 to 9 and then reset to
    0 on the tenth pulse.
3   Check the circuit using the pulser to supply clock pulses and using a logic clip (monitor)
    observe the state of each pin of the integrated circuits, during circuit operation.

4 Simulate the faults indicated in the table below. Observe how each fault affects the system and complete the symptoms column.

FAULT	SYMPTOM	FAULT	SYMPTOM
1 Enable input open circuit		5 Pins 3 and 4, 4511 disconnected	
2 Pin 12, 4011 disconnected		6 Pin 3 LED display disconnected	
3 R input (pin 15) disconnected		7 R input 4520 short circuit to 0 V line	
4 Pin 1, 4511 disconnected			

5 For each fault use the logic clip pulser, and logic probe to monitor the faulty circuit in operation and gain experience of the various techniques used for fault diagnosis.

ance meter has two probes one is placed at point A and one at point B. A low resistance is obtained due to the short circuit. Resistance between points A and C will be lower still and between points A and D will be higher than between A and C, thus showing that C is the position of the short circuit. Most instruments provide an audio output that rises in frequency as the resistance reduces, allowing the operator to trace the fault without having to look at the resistance reading.

Practical Investigation 24 illustrates this technique.

The Millivoltmeter

A printed circuit board may develop a partial short circuit between the tracks. This can be caused by a build up of dirt (particularly if it contains carbon) on a faulty component. To detect this the voltage drop between the tracks can be measured and the partial short located. Since direct current is used capacitors do not present problems and should the fault be a short circuit capacitor this will be discovered. The very low voltages (mV) that are involved mean that fault detection using this method can be tricky.

Practical Investigation 25 will give you some experience of this.

Diagnosing a Fault

After this very brief introduction to the test instruments you are now in a position to diagnose a fault on digital equipment. It is important to remember that fault location is a logical process involving rational thought processes, so prior to getting down to the actual testing of the circuit boards themselves it is worth observing the following procedure.

1 Establish what is wrong first! In order to diagnose a fault the exact symptoms must be established so that a preliminary diagnosis can be made. This may involve questioning the person who requires the repair. It will probably mean that you have to test the equipment following the manufacturer's test procedure.

2 Once the nature of the problem has been identified and the equipment is officially declared faulty, it can now be opened up and subjected to a visual inspection. Examine each board carefully looking for:
   (a) Cracked printed circuit boards.
   (b) Loose or missing ICs (they can jump out of their sockets if the equipment is dropped).
   (c) Disconnected leads (particularly ribbon cables that plug into pcb sockets).
   (d) pcbs loose in their edge connectors.

3 Evidence of excessive current flow, e.g.

burnt or charred areas on the board or blackened components; (a sensitive nose helps here).

If nothing is discovered during this inspection then it is time to use the instruments and procedures previously outlined. Remember to test and eliminate each section before moving on to the next area.

Practical Investigation 26 allows you to build a counter and then introduce faults that can be diagnosed using the methods previously discussed.

Other types of Test Equipment

Many digital systems operate at very high speeds. This means that the binary information or data is switched and shifted around at an incredible speed. To monitor these rapidly changing signals is very difficult. It can sometimes be achieved by slowing down the clock rate but this is not always possible, nor does it give a true picture of what is happening in terms of the real time operation of the system. Specialist instruments have now been developed that help to overcome these problems, such equipment although quite complex and expensive is worth considering.

Signature Analysis

A digital system like a microprocessor will produce data in a stream of binary pulses. This bit stream can be sampled and a code word generated that represents the sampled data. This word is in effect the signature of the data stream. By obtaining and recording the signature code word at various points in a system a bank of signatures can be built up. This has to be done initially on a fault free system and the information used later if the equipment fails. This is similar to the technique of indicating voltage levels as waveform shapes on the circuit diagram of analogue equipment. With digital systems the signatures can be written on the block diagram as shown in Fig. 7.8.

The signature analyser itself usually requires the following inputs from the circuit under test: clock, stop, start, data. There are facilities for

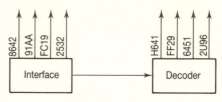

Fig. 7.8   Signatures

selecting the positive or negative edge of the pulses. The actual signature is displayed as a four digit alphanumeric code word. This is not a hexadecimal or ASCII code word it is simply a four character word that represents the signature.

Logic Analyser

This is an instrument that is provided with 8, 16 or 32 channels according to the model. It is able to monitor and record the logic levels at many points in a circuit simultaneously. This information is stored digitally. It can be displayed on a cathode ray tube as waveforms showing all the channels' data simultaneously in the form of a timing diagram as shown in Fig. 7.9.

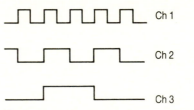

Fig. 7.9   Waveform display on a logic analyser

Alternatively the stored information can be displayed as code words (binary, octal, ASCII or hexadecimal) showing the monitored points in relation to each clock pulse, as indicated in Fig. 7.10.

Clk pulse	Data
01	0055
02	0AFA
03	4A2H
04	3FFF
05	0004
06	13AF
07	002H

Fig. 7.10   Data listed on a logic analyser

There are many different types of logic analysers available and most enable purpose made pods to be purchased that allow easy connection to modern microprocessors.

A final word about fault diagnosis

The ability to fault find quickly and accurately is only acquired by experience. A good car mechanic is able to diagnose and repair an engine by virtue of logical deduction and experience. Simulated circuit fault diagnosis cannot provide the necessary experience. In fact it can be downright misleading! It is just as difficult creating a high resistance fault in a component as it is to create an artifically worn out bearing in an engine. This final chapter has merely served to introduce the various test instruments and give very basic instruction on how to use them, the rest is really up to you.

Fault Diagnosis Review

1 Faults on digital circuits may occur in the integrated circuit, discrete component or the circuit board.

2 Before dismantling the equipment run a diagnostic check to determine the symptoms that may help to indicate the fault.

3 Use your senses of sight and smell to help locate problem areas on the circuit board, i.e. overheating, breaks in wires or tracks.

4 A logic probe is used to indicate the logic state at points in a circuit.

5 A logic pulser is a probe that injects precise binary pulses into a circuit for testing purposes.

6 A logic clip or monitor is a multi-display logic indicator that clips directly to an integrated circuit and indicates the logic state of each pin.

7 A current tracer allows the tiny currents flowing in the printed circuit board tracks to be followed so that the exact point of a short circuit may be found.

8 A milliohm meter allows the very low resistance of a short circuit to be traced.

9 A millivoltmeter can be used to measure the volt drop along the printed circuit board tracks enabling a partial short circuit to be located.

10 Digital systems that utilize high speed switching circuits and large amounts of data like microprocessors present special fault location problems.

11 Signature analysers and logic analysers are specialised instruments for locating faults on microprocessor equipment.

8

CMOS Data sheets

4001B Quad 2 input NOR

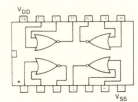

4002B Dual 4 input NOR

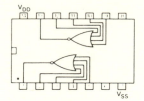

4008B 4 bit full adder

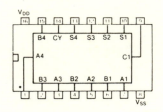

4011B Quad 2 input NAND

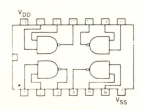

4012B Dual 4 input NAND

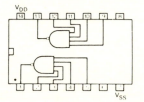

4013B Dual D-type flip-flop

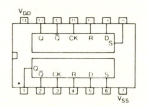

4014B 8 bit shift register

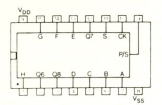

4015B Dual 4 bit shift register

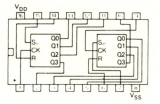

4017B Decade counter divider

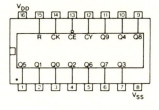

4020B 14 bit binary counter

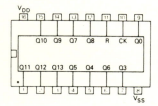

4021B 8 bit-shift register

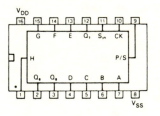

4023B Triple 3 input NAND

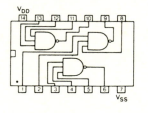

4024B Seven stage ripple counter

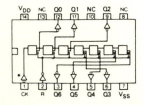

4025B Triple 3 input NOR

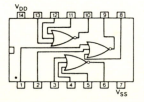

4027B Dual J.K. flip-flop

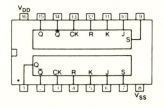

4028B BCD — decimal/binary-
octal decoder

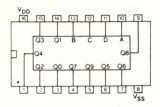

4029B Presettable binary/BCD
up/down counter

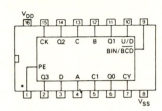

4035B 4 bit parallel — in/parallel —
out shift register

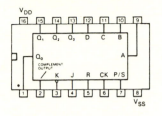

4040B 12 bit binary counter

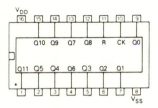

4042B Quad 'D' latch

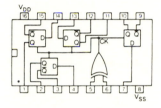

4043B Quad R/S latch with 3-state
outputs "NOR"

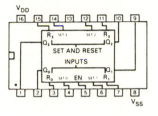

4044B Quad R/S latch with 3-state
outputs "NAND"

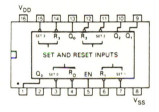

4047B Monostable/Astable
multivibrator

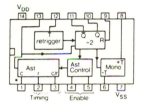

4049UB Hex inverter — buffer

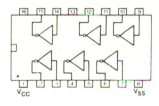

4050B Hex buffer

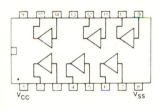

4051B 8 input analogue
multiplexer

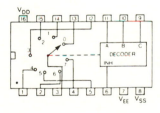

4052B Dual 4 input
analogue multiplexer

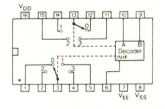

4053B Triple 2 input analogue
multiplexer

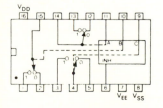

4060B 14 bit binary counter

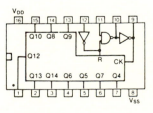

4068B 8-input NAND gate

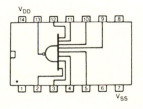

4069UB Hex inverter

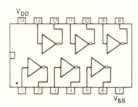

4070B Quad exclusive OR

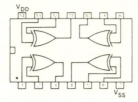

4071B Quad 2 input OR

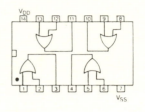

4072B Dual 4-input OR gate

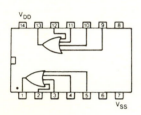

4073B Triple 3 input AND

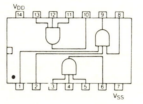

4075B Triple 3 input OR

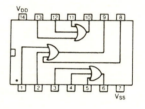

4078B 8 input NOR

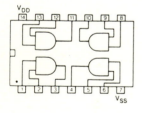

4077B Quad 2 input Exclusive "NOR" gate

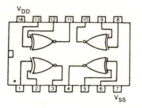

4076B Quad D type register

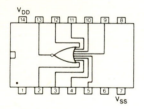

4081B Quad 2 input AND

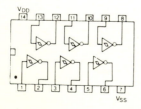

4160B Synchronous programmable 4 bit decade counter with asynchronous clear

4161B Synchronous programmable 4 bit binary counter with asynchronous clear

4162B Synchronous programmable 4 bit decade counter with synchronous clear

4163B Synchronous programmable 4 bit binary counter with synchronous clear

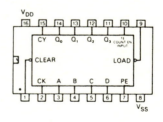

40106B Hex inverting schmitt

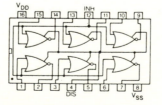

4502B Strobed Hex inverter/buffer

4510B BCD up/down counter

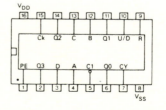

4511B BCD - 7 segment latch/ decoder/driver

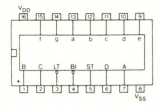

4512B 8 channel data selector

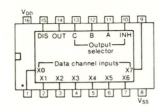

4513B BCD to seven segment latch/ decoder/driver

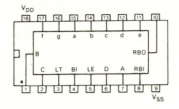

4516B Binary up/down counter

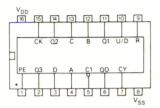

4518B Dual BCD up-counter

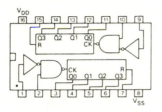

4519B Quad 2 input multiplexer

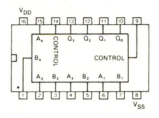

4520B Dual 4 bit binary counter

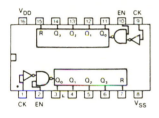

4528B Dual resettable monostable

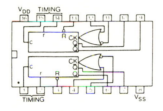

4529B Dual 4-channel analog data selector three state outputs

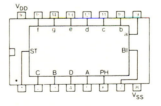

4532B 8 bit priority encoder

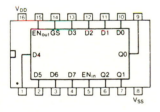

4539B Dual 4 channel data selector/ multiplexer

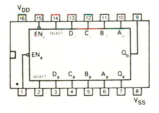

4543B BCD-to-seven segment latch/decoder/driver

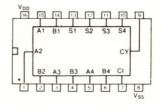

4551B Quad 2-input analog multiplexer/ demultiplexer

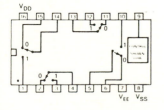

4553B Three-digit BCD counter

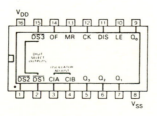

4560B Natural BCD adder

9

TTL Data sheets

00 Quadruple 2-input NAND gate

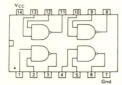

01 Quadruple 2-input NAND gate with open collector output

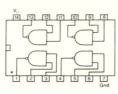

02 Quadruple 2-input NOR gate

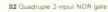
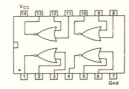

03 Quadruple 2-input NAND gate open collector inputs

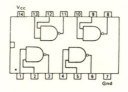

04 Hex inverter. A 74HC004 (un-buffered version) is also available.

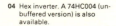
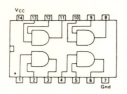

05 Hex inverter-open collector outputs

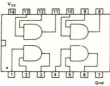

06 Hex inverter with high voltage open collector output

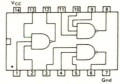

07 Hex driver with open collector output

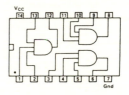

08 Quadruple 2-input AND gate

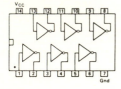

09 Quad 2-input AND gate-open collector outputs

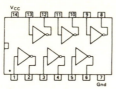

10 Triple 3-input NAND gate

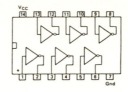

11 Triple 3-input AND gate

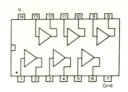

13 Dual 4-input NAND gate Schmitt trigger

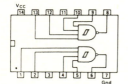

14 Hex Schmitt Trigger

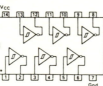

15 Triple 3-input AND gate — open collector outputs

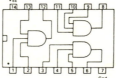

16 Hex Inverter with open collector output

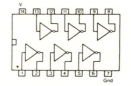

20 Dual 4-input NAND gate

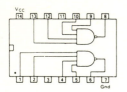

21 Dual 4-input AND gate

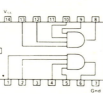

22 Dual 4-input NAND gate — open collector outputs

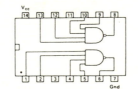

25 Dual 4-input NOR gate with strobe

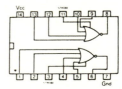

26 Quad 2-input NAND buffer-open collector outputs

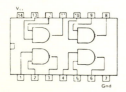

27 Triple 3-input NOR gate

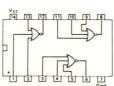

28 Quad 2-input NOR buffer

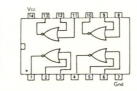

30 8-input NAND gate

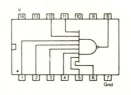

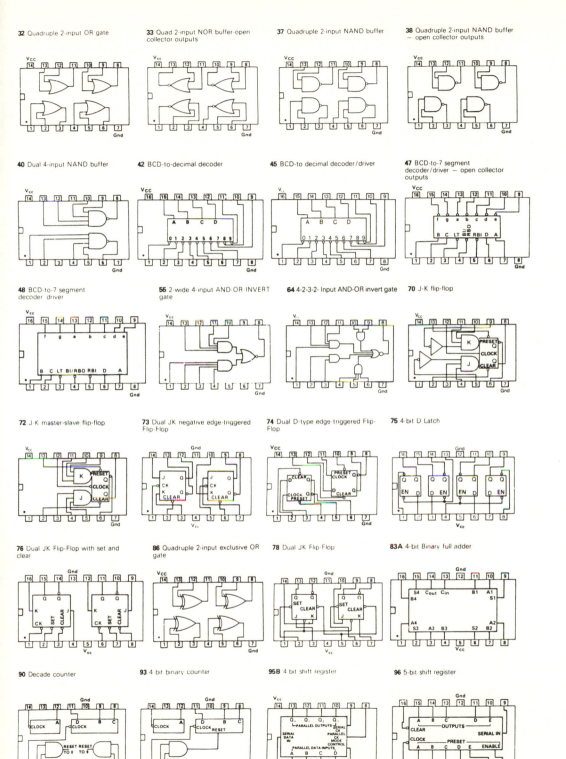

107 Dual JK Flip-Flop

109 Dual JK positive edge-triggered Flip Flop

112 Dual JK edge triggered flip flop

113 Dual JK negative edge-triggered Flip-Flop

114 Dual JK negative edge-triggered Flip-Flop

125 Quad 3-state buffer (active low enable)

126 Quad 3-state buffer (active high enable)

133 13-input NAND gate

137 3-line to 8-line Decoder/Demultiplexer with address latches

138 3 to 8 line Decoder/Multiplexer

151 1 of 8 Data Selector/Multiplexer

153 Dual 4 line to 1-line Data Selectors/Multiplexers

155 Dual 1 of 4 Decoder/Demultiplexer

156 Dual 1-of-4 Decoder/Demultiplexer with open collector outputs

157 Quad 2 to 1-line Data Selectors/Multiplexers

158 Quad 2 to 1-line Data selectors/Multiplexers with inverted outputs

160 BCD decade counter – asynchronous reset

161 Binary counter – asynchronous reset

162 BCD counter – synchronous reset

163 Binary counter – synchronous reset

164 Serial-in parallel-out shift register

165 8-bit parallel to serial converter

166 8-bit shift register

169 4-Stage synchronous bidirectional counter

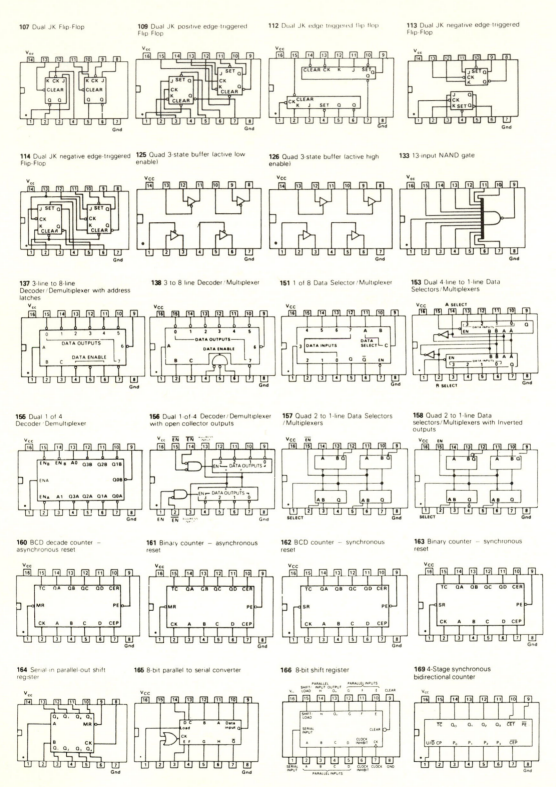

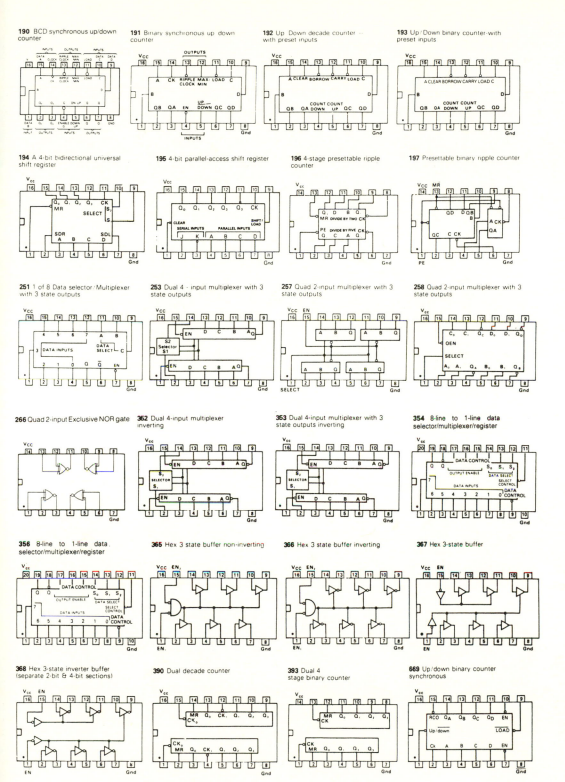

190 BCD synchronous up/down counter

191 Binary synchronous up/down counter

192 Up/Down decade counter — with preset inputs

193 Up/Down binary counter-with preset inputs

194 A 4-bit bidirectional universal shift register

195 4-bit parallel-access shift register

196 4-stage presettable ripple counter

197 Presettable binary ripple counter

251 1 of 8 Data selector/Multiplexer with 3 state outputs

253 Dual 4-input multiplexer with 3 state outputs

257 Quad 2-input multiplexer with 3 state outputs

258 Quad 2-input multiplexer with 3 state outputs

266 Quad 2-input Exclusive NOR gate

362 Dual 4-input multiplexer inverting

353 Dual 4-input multiplexer with 3 state outputs inverting

354 8-line to 1-line data selector/multiplexer/register

356 8-line to 1-line data selector/multiplexer/register

365 Hex 3 state buffer non-inverting

366 Hex 3 state buffer inverting

367 Hex 3-state buffer

368 Hex 3-state inverter buffer (separate 2-bit & 4-bit sections)

390 Dual decade counter

393 Dual 4 stage binary counter

669 Up/down binary counter synchronous

10

Analogue to digital converter integrated circuits

Sample-and-Hold Circuits

LF398H

Supplied to RS by National Semiconductor

A monolithic sample-and-hold integrated circuit implemented using BI-F.E.T. technology to obtain high d.c. accuracy, fast signal acquisition and low droop rate. Input offset voltage typically 0·5 mV, acquisition time < 10 μs. Input characteristics do not change during hold mode. Logic inputs. T.T.L./C-MOS compatible. Supply voltage range: ± 5 to ± 18V. Operating temperature range: 0°C to + 70°C. TO99 metal can package.

TOP VIEW

High Speed Precision AD585AQ

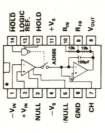

TOP VIEW

technical specification	
Supply voltage	+ 5, -12 to ± 18 V
Supply current	10 mA
Acquisition time (10 V step to 0·01%)	3 μs
Aperture jitter	0·5 ns
Droop rate (max.)	1 mV/ms
Sample to hold offset (max.)	3 mV
Small signal gain bandwidth	2 MHz
Offset voltage (max.)	2 mV

Supplied to RS by Analog Devices

The AD585 is a monolithic sample-and-hold circuit consisting of a high performance operational amplifier in series with an ultra-low leakage analog switch and a F.E.T. input integrating amplifier. An internal holding capacitor and connections to the internal feedback resistors, completes the sample and hold.

With the switch closed, the AD585 functions like a standard op-amp; any feedback network may be connected around the device to control gain and frequency response. With the switch open, the capacitor holds the output at its previous level. The AD585 offers a combination of fast acquisition time (3·0 μs to 0·01%) and low offset step (3 mV) making the device **suitable for high speed 12-bit data acquisition systems.**

A-D Converters

8-bit
ZN439E-7
ZN439E-8

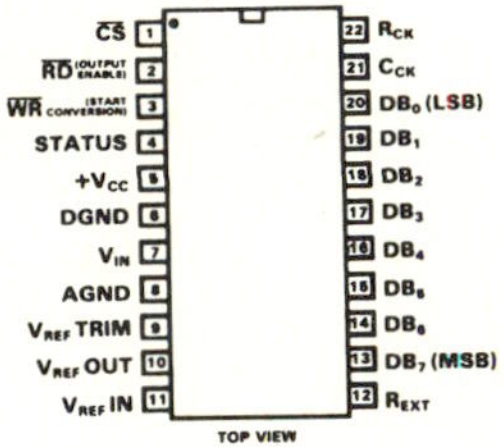

TOP VIEW

Pin labels: C̄S̄, RD (OUTPUT ENABLE), W̄R̄ (CONVERT START), STATUS, +Vcc, DGND, Vin, AGND, VREF TRIM, VREF OUT, VREF IN / RCK, CCK, DB0 (LSB), DB1, DB2, DB3, DB4, DB5, DB6, DB7 (MSB), REXT

The ZN439 is a complete 8-bit A to D converter. All the active circuitry is on-chip including the clock generator, trimmable 2·5 V bandgap reference, control logic and double buffered latches with 3-state outputs. Just 3 inputs are used to control all the converter operations and the output latch configuration allows data to be read at any time. Two versions are available; the -8 type should be used where greater accuracy is required.
22-pin d.i.l. plastic package.

technical specification

	ZN439E-7	ZN439E-8
Supply voltage	5 V ± 10%	
Supply current	45 mA max.	
Resolution	8 bits	
Linearity error	±1 LSB	±½ LSB
Differential linearity error	±1 LSB	±¾ LSB
Conversion time	5 µs max.	

8-bit, High-Speed 'Flash'
MC10319P

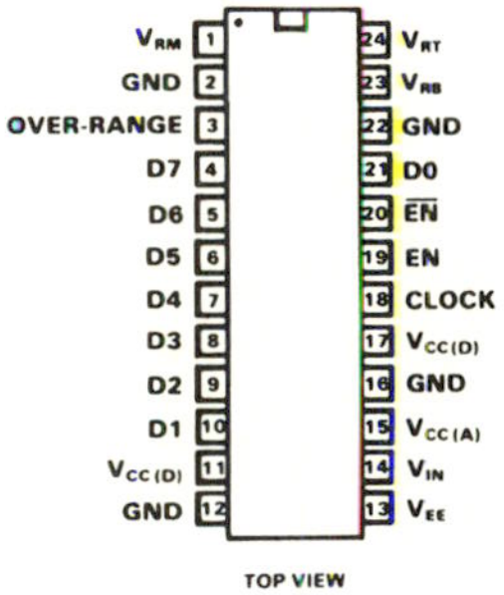

TOP VIEW

Pin labels: V_{RT}, V_{RB}, GND, DO, ĒN, CLOCK, $V_{CC(D)}$, GND, $V_{CC(A)}$, V_{IN}, V_{EE}, GND, D1, D2, D3, D4, D5, D6, D7, GND, OVER-RANGE, V_{RM}

An 8-bit high-speed parallel flash A to D converter which employs an internal grey code structure to eliminate large output errors on fast slewing input signals.
The MC10319 is fully T.T.L. compatible with three-state T.T.L. outputs which enable direct drive of a data bus or a common I/O memory. The over-range bit is always active in order to sense an over-range condition or ease the interconnection of a pair of devices to produce a 9-bit A to D converter.
Applications include video/TV encoding and high-speed instrumentation.

technical specification

Supply voltage, V_{CC}(A), V_{CC} (D)	+5·0 V (typ.)
Supply voltage, V_{EE}	−5·0 V (typ.)
Digital input voltages	0 V to +5 V d.c.
Analog input (pin 14)	−2·1 V to +2·1 V d.c.
Clock pulse width:	
High	20 ns (typ.)
Low	20 ns (typ.)
Clock frequency	25 MHz (max.)
Resolution	8 bits (max.)
Monotonocity	Guaranteed
Diff non-linearity	±1 LSB (max.)
Diff gain	1% (typ.)

8-bit Ultra High Speed
AD9002AD

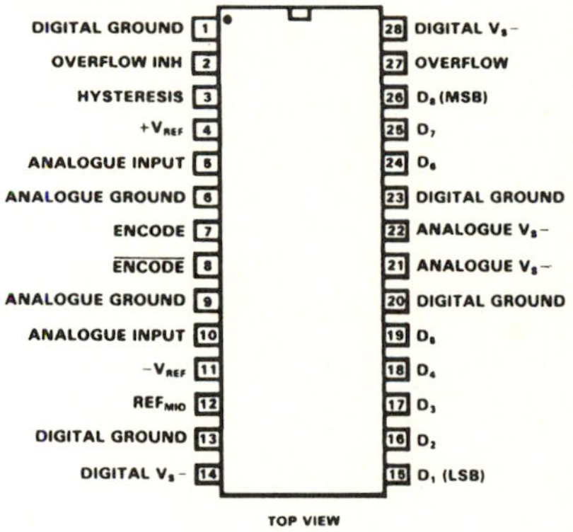

TOP VIEW

Pin labels: DIGITAL GROUND, OVERFLOW INH, HYSTERESIS, +VREF, ANALOGUE INPUT, ANALOGUE GROUND, ENCODE, ENCODE, ANALOGUE INPUT, ANALOGUE GROUND, −VREF, REFMID, DIGITAL GROUND VL / DIGITAL VL, OVERFLOW, D8 (MSB), D7, D6, DIGITAL GROUND, ANALOGUE VL, ANALOGUE VL, DIGITAL GROUND, D5, D4, D3, D2, D1 (LSB)

Supplied to RS by Analog Devices

An exceptionally fast analogue to digital converter with sampling rates in excess of 150 megasamples per second. Comparator design on the AD9002AD is very advanced and features an extremely wide small signal bandwidth of 115 MHz, and this allows very accurate acquisition of high-speed pulse inputs, without needing an external track-and-hold.
Comparator output comprises 256 parallel stages with outputs decoded to drive the ECL compatible output latches; the comparator output decoding minimises false codes. An external hysteresis pin is provided to optimise comparator sensitivity. Power dissipation is only 750 mW allowing use over the full industrial temperature range.
Supplied in a 28-pin d.i.l. plastic package.

technical specification

Resolution	8 bits
Differential and integral	
Linearity at 25°C (d.c.)	0·6 LSB typ.
Analogue Input	
Voltage range	−2·1 to +0·1 V
Bias current	60 µA typ.
Resistance	20 kΩ typ.
Reference Input	
Differential ref. voltage	2 V
Dynamic performance	
Conversion time (+1 clock)	4 ns
Conversion rate	150 MHz
Aperture delay	1·3 ns
Jitter	15 ps
Output delay	3·7 ns typ.
Transient response	6 ns
Signal-to-noise ratio	46 dB
Temperature range	−25°C to +85°C
Power supply	
Voltage	−4·94 to −5·46 V
	−5·2 V typ.
Current	145 mA at −5·2 V

11 Display devices

Seven segment LED displays

0·3in

W. 10
H. 19
D. (ex. pins) 6

PIN CONNECTIONS (TOP VIEW)

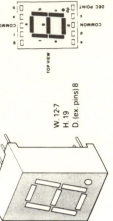

Gallium arsenide phosphide 7-segment displays in 14-pin d.i.l. packages. Common anode version incorporates a left-hand decimal point, common cathode a right-hand. Series resistors must be used with each segment to limit I$_F$.

0·3in low current

W. 7·62 H. 12·7 D. 5·08
Pin spacing 2·54 Row spacing 5·08

PIN CONNECTIONS (TOP VIEW)

COMMON 1

0·3 in high, 7-segment, low current high efficiency red L.E.D. displays. Available in **common anode** and **common cathode** configurations, both with right-hand decimal points. Displays directly compatible with T.T.L./C-MOS integrated circuits. Displays are categorised for luminous intensity and are designed for optimium on/off contrast.

0·56in low current

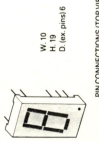

W. 12·6 H. 17 D. 8
Pin spacing 2·54 Row spacing 15·24

0·56 in high, 7-segment, low current high efficiency **Red** LED displays available in **common anode** and **common cathode** configurations, both with right-hand decimal points. Displays are directly compatible with TTL/C-MOS integrated circuits. Luminous intensity and forward voltage tested at 3mA to assure consistent brightness at TTL output current levels.

0·5in

TOP VIEW

W. 12·7
H. 19
D. (ex. pins) 8

0·5 in high, 7-segment red L.E.D. display available in both **common anode** and **common cathode** versions. The connections are along the top and bottom edge simplifying wiring in multi-digit applications - especially multiplexed display. Both types incorporate right-hand decimal points. Series resistors must be used with each segment and decimal point to limit the forward current.

1in

W. 22·5
H. 33
D. (ex. pins) 8·5

COMMON
NO PIN
NO PIN
COMMON

14 COMMON
13 b
12 NO PIN
11 c
10 c
9 DEC. POINT
8 d

Large 1 in high red 7-segment L.E.D. displays available in both **common anode** and **common cathode** versions. T.T.L. compatible. Series resistor must be used with each segment to limit I$_F$.

2·24in

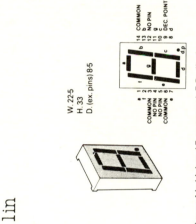

COMMON
COMMON

H. 70
W. 48
D. 12
Pin spacing 2·54
Row spacing 60

2·24 in, high efficiency red, 7-segment L.E.D. displays available in both **common anode** and **common cathode** configurations, both with right-hand decimal point. Each segment contains 4 chips, in series (decimal point has 2 chips) and is therefore recommended for 12 V operation. Series resistor must be used with each segment to limit I$_F$. Displays are designed for maximum on/off contrast. Connections are along the top and bottom edge

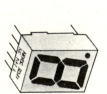

additional data - LED displays

	0.3in Red	0.3in Green	0.43in low current	0.43in	0.5in	0.56in	0.56in low current	0.8in	1in	2.24in	2.24in Star burst	Dot Matrix 0.7in	1-16in	2in
Operating temperature range	-40°C to +85°C	-40°C to +85°C	-40°C to +85°C	-40°C to +85°C	-25°C to +85°C	-40°C to +85°C	-40°C to +85°C	-40°C to +85°C	-20°C to +70°C	-40°C to +85°C	-40°C to +85°C	-40°C to +85°C	-40°C to +85°C	-40°C to +85°C
Ratings per segment PD(max) at 25°C	60mW	75mW	52mW	105mW	50mW	105mW	52mW	50mW	89mW	*420mW 210mW (D.P.)	60mW	60mW	60mW	60mW
IF	10mA typ. 30mA max.	10mA typ. 30mA max.	2mA typ. 15mA max.	20mA typ. 60mA max.	20mA typ.	10mA typ.	2mA typ.	20mA typ.	20mA typ.	20mA typ.	30mA	20mA max.	10mA typ.	10mA typ.
Derating above 25°C	0.5mA/°C	0.4mA/°C	0.4mA/°C† 0.6mA/°C‡	0.4mA/°C	25mA max. 0.3mA/°C	60mA max. 0.4mA/°C	15mA max. 0.4mA/°C† 0.6mA/°C‡	25mA max. 0.33mA/°C	25mA max. 0.45mA/°C	30mA max. 0.45mA/°C	8.4V typ. at 20mA 4.2V typ. for D.P. 5V max.		20mA max. 0.4mA/°C	20mA max. 0.4mA/°C
VF	1.6V typ. at 10mA	2.2V typ. at 10mA	1.6V typ. at 2mA	2.1V typ. at 20mA	1.7V typ. at 20mA	2.1V typ. at 10mA	1.6V typ. at 2mA	1.7V typ. at 20mA	1.7V for D.P. 3.3V typ. at 20mA	8.4V typ. at 20mA 4.2V typ. for D.P. 20V		2.1V typ. at 10mA	2.1V typ. at 10mA	2.1V typ. at 10mA
VR max.	6V max.	6V max.	3.25V min.	3V min.	3V max.	3V min.	3V min.	3V max.	6V max. 5V for D.P.	10V for D.P.	5V max.	5V max.	5V max.	5V max.
Luminous intensity typ. at IF typ. (digit)	2.5mcd	2.0mcd	*270µcd	*300µcd	*28mcd	*28mcd	*370µcd	2.2mcd	3.6mcd	6mcd	3mcd	2mcd	2mcd	2.2mcd

† Derate above 65°C at 0.4mA/°C
‡ Derate above 78°C at 0.6mA/°C
* Luminous intensity per segment
* Power dissipation typical

Starburst and dot matrix displays

2.24in star burst

H. 70
W. 48
D. 10
Pin spacing 2.54
Row spacing 60

2.24 in, high efficiency red 'star burst' L.E.D. displays, available in common anode and cathode configurations both with right hand decimal points.

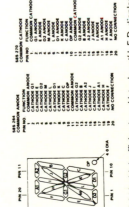

dot matrix

Large high efficiency red 5 × 7 dot matrix L.E.D. displays with X-Y select. Available in two matrix configurations being either **common anode** row or **common cathode** row. Designed for maximum on/off contrast. Displays are categorised for luminous intensity matching. Ideal applications include electronic instrumentation, computer peripherals, etc.

17.33mm (0.68in)

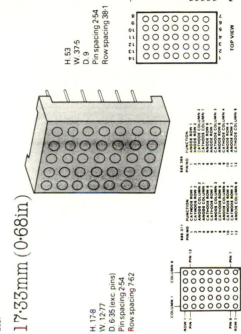

H. 17.8
W. 12.77
D. 6.35 (exc. pins)
Pin spacing 2.54
Row spacing 7.62

50.8mm (2in)

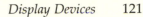

H. 53
W. 37.5
D. 9
Pin spacing 2.54
Row spacing 38.1

589-165

PIN NO	FUNCTION
1	ANODE ROW 5
2	ANODE ROW 7
3	ANODE COLUMN 2
4	ANODE COLUMN 3
5	ANODE ROW 4
6	CATHODE COLUMN 5
7	ANODE ROW 6
8	ANODE ROW 3
9	ANODE ROW 1
10	CATHODE COLUMN 4
11	CATHODE COLUMN 3
12	ANODE ROW 4
13	CATHODE COLUMN 1
14	ANODE ROW 2

589-171

PIN NO	FUNCTION
1	CATHODE ROW 5
2	CATHODE ROW 7
3	ANODE COLUMN 2
4	ANODE COLUMN 3
5	CATHODE ROW 4
6	ANODE COLUMN 5
7	CATHODE ROW 6
8	CATHODE ROW 3
9	CATHODE ROW 1
10	ANODE COLUMN 4
11	ANODE COLUMN 3
12	CATHODE ROW 4
13	ANODE COLUMN 1
14	CATHODE ROW 2

NOTE: PINS 4 AND 11 ARE INTERNALLY CONNECTED
PINS 5 AND 12 ARE INTERNALLY CONNECTED

Liquid crystal displays

alphanumeric dot matrix L.C.D.

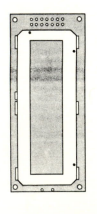

format character × line	dimensions L	W	D	character viewing area L	W
16 × 1	85	36	9.3	59.4	8.7
16 × 2	85	36	9.3	59.4	12.3
24 × 2	118	36	9.3	89.4	12.3
32 × 2	174.5	31	11	141.0	16.5
40 × 2	182	33.5	9.3	149.4	12.3
40 × 4	215	60	17.5	149.4	27.25

Intelligent, alphanumeric, dot matrix modules with integral C-MOS microprocessor and L.C.D. display drivers. The modules utilise a 5 × 7 dot matrix format, with cursor, and are capable of displaying the full ASCII character set plus up to eight additional user programmable custom symbols.

Features include:
- Single power supply +5 V, 2 mA.
- T.T.L. and +5 V C-MOS compatible.
- Direct interface to any 4 bit or 8 bit microprocessor bus.
- Wide, adjustable, viewing angle.
- Powerful instructions save lines of coding:
  Display - clear, shift, on/off, read, write.
  Cursor - home, on/off, shift, set direction.
- Scroll left, right or alternate with entire line replacement.

alphanumeric dot matrix LCD module with EL back lighting

Format Character × Line	Dimensions L	W	D	Character view area W	H	Character size W	H
8 × 2	58	32	12	35	15	2.45	3.8
16 × 1	85	36	12	63.5	15.8	3.15	7.9
16 × 2	85	36	12	63.5	15.8	3.15	4.45
24 × 2	118	36	13	93.5	15.8	3.15	4.45
40 × 2	182	33.5	13	154.5	15.8	3.15	4.45
40 × 4	215	60	17.5	160	34	3.2	4.85

A range of intelligent alphanumeric dot matrix displays with integral green electroluminescent back lighting facility. The EL back lighting provides a low power solution to the problem of illuminating LCD's. An inverter is required to power this facility. See selection table below.

module	display voltage	inverter type	back light Vf	Fo
8 × 2	±5V	585-062	82 Vrms	410Hz
16 × 1	±5V	585-062	82 Vrms	410Hz
16 × 2	±5V	585-062	82 Vrms	410Hz
24 × 2	±5V	585-062	82 Vrms	410Hz
40 × 2	±5V	585-062	82 Vrms	410Hz
40 × 4	±5V	585-078	75 Vrms	750Hz

The modules utilise a 5 × 7 dot matrix format (except for the 16 × 1 module, which uses a 5 × 10), with cursor, and are capable of displaying 160 different alphanumeric characters and symbols. The user can also produce any character pattern using the on board RAM facility.

Index